# Handbook of Disruptive Technologies

This handbook addresses how smart operations, management, and healthcare can be used to detect and analyse supply chain problems, business problems, and diseases. It also discusses the underlying methodologies and related security concerns.

*Handbook of Disruptive Technologies: Operations, Business, Management, and Healthcare* provides updated and timely insights, unique approaches or frameworks, practical applications, and case studies. It provides in-depth knowledge of disruptive technologies and applies data analytics to different areas to solve complex situations. The handbook uses data analytics in decision-making and policy framing and identifies practices and applications of disruptive technologies that can assist organizations in gaining a competitive advantage. Valuable insights, innovative approaches, practical applications, and case studies can be found in this handbook as it effectively consolidates and synthesizes information to provide a comprehensive understanding of the subject matter.

This book is tailored for researchers, academics, and professionals seeking to deepen their knowledge in the field.

**Advancements in Intelligent and Sustainable Technologies and Systems**

*Series Editor:*
Ajay Kumar

This book series aims to provide a platform for academicians, researchers, professionals, and individuals to participate and to provide novel systematic theoretical, experimental, computational work in the form of edited text or reference books, monographs in the area of intelligent and sustainable technologies and systems from engineering, management, applied science, healthcare, etc. domains. This book series will educate and inform the readers with a comprehensive overview of advancements in intelligent and sustainable techniques and systems with novel intelligent tools and algorithms to move industries from different domains from a data-centric community to a sustainable world. The book series covers ideas and innovations to help the research community and professionals to understand fundamentals, opportunities, challenges, future outlook, layout, lifecycle, and framework of intelligent and sustainable technologies and systems for different sectors. It serves as a guide for computer science, mechanical, manufacturing, electrical, electronics, civil, automobile, industrial engineering, biomedical, healthcare, and management professionals.

*If you are interested in writing or editing a book for the series or would like more information, please contact Cindy Carelli,* cindy.carelli@taylorandfrancis.com

Handbook of Intelligent and Sustainable Smart Dentistry
Nature and Bio-Inspired Approaches, Processes, Materials, and Manufacturing
*Edited by Ajay Kumar, Namrata Dogra, Sarita, Surbhi Bhatia, M S Sidhu*

Handbook of Intelligent and Sustainable Manufacturing
Tools, Principles, and Strategies
*Edited by Ajay Kumar, Parveen Kumar, Yang Liu, and Rakesh Kumar*

Advances in Sustainable Biomaterials
Bioprocessing 4.0, Characterizations, and Applications
*Edited by Ajay Kumar, D.K. Rajak, Parveen Kumar, and Ashwini Kumar*

Manufacturing Strategies and Systems
Technologies, Processes, and Machine Tools
*Edited by Ajay Kumar, Parveen Kumar, Ashish Kumar Srivastava, and Lokesh Saharan*

Handbook of AI in Engineering Applications
Tools, Techniques, and Algorithms
*Edited by Ajay Kumar, Sangeeta Rani, Krishna Dev Kumar, and Manish Jain*

Handbook of Deep Learning Models for Healthcare Data Processing
Disease Prediction, Analysis, and Applications
*Edited by Ajay Kumar, Deepak Dembla, Seema Tinker, and Surbhi Bhatia Khan*

Handbook of Disruptive Technologies
Operations, Business, Management, and Healthcare
*Edited by Chand Prakash, Neha Gupta, and Ajay Kumar*

# Handbook of Disruptive Technologies

## Operations, Business, Management, and Healthcare

Edited by Chand Prakash,
Neha Gupta, and Ajay Kumar

CRC Press
Taylor & Francis Group
Boca Raton London New York

CRC Press is an imprint of the
Taylor & Francis Group, an **informa** business

Designed cover image: Shutterstock – Vasin Lee

First edition published 2025
by CRC Press
2385 NW Executive Center Drive, Suite 320, Boca Raton FL 33431

and by CRC Press
4 Park Square, Milton Park, Abingdon, Oxon, OX14 4RN

*CRC Press is an imprint of Taylor & Francis Group, LLC*



ISBN: 978-1-032-69987-5 (hbk)
ISBN: 978-1-032-70094-6 (pbk)
ISBN: 978-1-032-70095-3 (ebk)

DOI: 10.1201/9781032700953

Typeset in Times
by codeMantra

# Contents

# Preface

The *Handbook of Disruptive Technologies: Operations, Business, Management, and Healthcare* explores cutting-edge technologies that revolutionize industries across the globe. Disruptive technologies like artificial intelligence (AI), blockchain, machine learning, Internet of Things (IoT), and digital twins are transforming traditional processes in operations, business, and healthcare, offering profound improvements in efficiency, innovation, and outcomes. This book highlights:

- New methodologies to integrate disruptive technologies in various sectors.
- Comprehensive insights from cross-disciplinary fields such as engineering, management, and healthcare.
- Advanced research in digital transformation, including Industry 4.0 and smart systems, shaping the future of business and healthcare practices.

This book consists of 13 chapters that describe perspectives of Disruptive Technologies in Operations, Business, Management, and Healthcare.

- Chapter "Embracing Disruption: Emerging Technologies Reshaping the Financial Services Sector" throws light on how digital transformation has redefined the financial service sectors, especially by the applicability of blockchain technology. The use of this technology has transformed the way banking transactions used to take place. The financial world is well-informed about the potential benefits of blockchain technology and that, going forward, it will have a substantial impact on financial services. Thus, the chapter highlights the importance and prominent uses of blockchain technology in the finance sector.
- Chapter "Disruptive Technologies for Energy Efficiency in Industries" presents an overview of the existing disruptive technologies used to enhance energy efficiency in industrial sector.
- Chapter "Artificial Intelligence as Disruptive Technology in Human Resource Management" focuses on the scenario that demonstrates how AI, a disruptive technology, has changed traditional human resource management (HRM). From this chapter, readers can gain an in-depth understanding of AI, such as what is AI and why it is called disruption in the beginning, and later how this disruptive AI transformed the entire HRM.
- Chapter "Data Quality Management Framework for Data Warehouse" highlights possible solution that can be applied to maintain and improve the quality of data in the data warehouse. The proposed solution includes various perspectives of data warehouse management that ranges from simple database user (operator) to database administrator (expert user).
- Chapter "An Empirical Investigation of Educators' Professional Expertise to Incorporate Artificial Intelligence–Based Tools into the Classroom in a Responsible and Ethical Manner: Approach Towards Intelligent-TPACK"

explains how professional educators can use AI tools in the classroom the right way: a pathway for building a responsible and ethical learning environment for students and teachers that will increase the learning outcome.

- Chapter "Marketing Automation Development as a Process of Digital Ecosystem Advancement" explores marketing automation as a transformative framework for businesses to streamline marketing strategies through technology-driven optimization. It emphasizes how automation enables deeper customer insights, personalization, and data-driven decision-making to enhance engagement. This chapter provides a conceptual and systematic analysis of marketing automation within the evolving digital ecosystem. Ultimately, it highlights its role in shaping modern marketing strategies by integrating digitization, customization, and contextualization.
- Chapter "Modeling of Invisible Cost Factors in Information Technology Offshore Outsourcing by ISM Approach" explores the Y2K problem as the catalyst for IT offshore outsourcing, which expanded globally to reduce costs and increase profitability. While offshore outsourcing gained traction due to advancements in technology and connectivity, it also introduced invisible costs related to language, culture, and service customization. Using Interpretive Structural Modeling (ISM) and MICMAC analysis, the research identifies relationships among these cost factors, categorizing them based on their influence. The findings help decision-makers prioritize critical cost factors and assess their impact on IT offshore outsourcing.
- Chapter "Analysis of Dynamics Between Cryptocurrencies and Foreign Exchange Rates in Indian Scenario" analyses the dynamics between selected cryptocurrencies and foreign exchange rates in the Indian context, highlighting their interrelationships and the presence of autocorrelation in regression models using over 1,300 data points from FY 2019 to FY 2023.
- Chapter "Intertwining Between Digitalization and Triple Bottom Line Sustainability in the Banking Sector: Sustainable Banking Digital Ecosystem and Key Dimensions" aims to examine the interconnection between banking digitalization and three dimensions of sustainability and pinpoints key indicators for measuring the sustainability impact of digitalization in banking through an extensive literature review.
- Chapter "Navigating Soft Technology Disruption: Transforming Accounting Education for Industry 4.0 and 5.0 Era" explores how advancements in soft technologies are reshaping the landscape of accounting education. It highlights the necessary curriculum changes and pedagogical strategies needed to prepare students for the evolving demands of Industry 4.0 and Industry 5.0, emphasizing skills like automation, data analytics, and AI integration in the accounting profession.
- Chapter "The Trend and Type Analysis of Various Disruptive Technologies in Healthcare" discusses trends in big data, blockchain, the IoT, clinical decision support systems, and cloud computing in healthcare. It also discusses the number of different types of available literature such as articles, reviews, book chapters, conference papers, and conference reviews based on the Scopus database.

- Chapter "Embracing Technological Advancements: The Transformation of Healthcare Services in the Digital Age" focuses on the need for digitalization and inclusion of digitalization in the Sustainable Development Agenda 2030, along with the restrictions and implications of digitalization in the healthcare sector, with particular emphasis on the Ayushman Bharat Digital Mission.
- Chapter "Unleashing the Potential of Machine Learning: Revolutionizing Loan Forecasting Using Logistic Regression" presents a machine-learning-based loan prediction model, utilizing logistic regression, to enhance lending efficiency and risk management in banking systems.

This book is intended for both the academia and the industry. Postgraduate students, Ph.D. students, and researchers in universities and institutions, who are involved in the areas of disruptive technologies in operations, business, management, and healthcare, will find this compilation useful.

The editors acknowledge the professional support received from CRC Press and express their gratitude for this opportunity.

Reader's observations, suggestions, and queries are welcome.

**Editors**
**Chand Prakash**
**Neha Gupta**
**Ajay Kumar**

# Acknowledgements

The editors are grateful to the CRC Press for showing interest in publishing this book in the buzz area of *Handbook of Disruptive Technologies: Operations, Business, Management, and Healthcare*. The editors express their personal admiration and gratitude to Ms. Cindy (Executive Editor) CRC Press, for giving consent to publish our work. He undoubtedly imparted great and adept experience in terms of systematic and methodical staff that helped the editors compile and finalize the manuscript. The editors also extend their gratitude to Ms. Kaitlyn Fisher CRC Press for supporting her during her tenure.

The editors wish to thank all the chapter authors who contributed their valuable research and experience to compile this volume. The chapter authors, corresponding author in particular, deserve special acknowledgements for bearing with the editors, who persistently kept bothering them with deadlines, and with their remarks.

Dr. Chand Prakash also wishes to express his gratitude to his parents, Mr. Ramphool Saini and Mrs. Radha Devi, for their true and endless support. They have raised him in a way that he can deliver this book along with co-editors. He cannot close these prefatory remarks without expressing his deep sense of gratitude and reverence to his life partner Mrs. Shivani Saini for her understanding, care, support, and encouragement to keep his morale high all the time. No magnitude of words can ever quantify the love and gratitude that he feels in thanking his sons Abhay and Praveen and daughter Suleka who are the world's best children.

Finally, the editors obligate this work to the divine creator and express their indebtedness to the "ALMIGHTY" for gifting them the power to yield their ideas and concepts into substantial manifestation. The editors believe that this book would enlighten the readers about the various dimensions of Disruptive Technologies in Operations, Business, Management, and Healthcare.

**Chand Prakash**
**Neha Gupta**
**Ajay Kumar**

# About the Editors

**Dr. Chand Prakash** is a distinguished academician with over 14 years of extensive experience in marketing, retail management, and consumer behaviour. He earned his doctorate in Management, focusing on critical issues in E-tailing and its impact on consumer behaviour. His academic journey is marked by a Master's degree in Management and Commerce, along with qualifying UGC NET in both disciplines and receiving a Junior Research Fellowship in Management. Dr. Prakash has made significant research contributions, authoring and co-authoring over 30 research papers published in prestigious journals, including ABDC, UGC Care, Scopus, and Web of Science (WoS)–listed publications. He has presented his findings at more than 50 national and international conferences, including those at renowned institutions like IITs and IIMs, and has received multiple Best Research Paper Awards. Additionally, he has published five cases at Case Centre UK. In the realm of editorial and academic contributions, Dr. Prakash has edited a book published by CRC Press and a special issue on "Integration of Analytics, Optimization, and Marketing for the Growth of Businesses in Emerging Economies" for the *Transnational Marketing Journal*, indexed in Scopus. He has conducted numerous seminars, workshops, and development programmes; chaired technical sessions at international conferences; and served as a reviewer for journals under Inderscience and IGI Global. He is frequently invited to deliver expert lectures at various institutions on his research and areas of expertise. Dr. Prakash's research interests are deeply rooted in quantitative studies, particularly structural equation modelling, and decision-making and data analysis using SPSS. His dedication to academia is further demonstrated by his successful mentorship of four doctoral research scholars, guiding them through their doctoral programmes to successful completions.

**Dr. Neha Gupta** is working as an associate professor in the Amity School of Business, Amity University, Noida, Uttar Pradesh, India. She obtained her Ph.D. from the Department of Statistics and Operations Research, Aligarh Muslim University, Aligarh. She is a gold medallist in M.Sc. in Operations Research. Her broad research interests include optimization modelling and decision sciences. She is a lifetime member of the Operational Research Society of India. She has more than 35 publications indexed in ABS, ABDC, SCIE, Scopus, and WoS. She has recently authored two books and edited two books for publishers like Taylor & Francis Group and Springer. She has recently edited two special issues for the *Journal of Pricing and Revenue Management*, Springer, and *Transnational Marketing Journal.*

**Dr. Ajay Kumar** is currently serving as a professor in Department of Mechanical Engineering in the School of Engineering and Technology, JECRC University, Jaipur, Rajasthan, India. He received his Ph.D. in the field of Advanced Manufacturing and Automation from Guru Jambheshwar University of Science and Technology, Hisar, India, after B.Tech. (Hons.) in Mechanical Engineering and M.Tech. (Distinction) in Manufacturing and Automation. His areas of research include biomedical engineering, incremental sheet forming, AI, sustainable materials, robotics and automation, additive manufacturing, mechatronics, smart manufacturing, Industry 4.0, industrial engineering systems, waste management, and optimization techniques. He has over 100 publications in international journals of repute including Scopus, WoS, and SCI indexed database and refereed international conferences. He has organized various national and international events including an International Conference on Mechatronics and Artificial Intelligence (ICMAI-2021) as conference chair. He is currently organizing international conference on Artificial Intelligence, Advanced Materials, and Mechatronics Systems (AIAMMS-2023) as conference chair. He has more than 20 national and international patents to his credit. He has supervised more than eight M.Tech., Ph.D. scholars, and numerous undergraduate projects/thesis. He has a total of 15 years of experience in teaching and research. He is a guest editor of many reputed journals. He has contributed to many international conferences/symposiums as a session chair, expert speaker, and member of the editorial board. He has won several proficiency awards during the course of his career, including merit awards, best teacher awards, and so on. He has also co-authored and co-edited more than 15 books and proceedings. He has also authored many in-house course notes, lab manuals, monographs, and invited chapters in books. He has organized a series of faculty development programmes, international conferences, workshops, and seminars for researchers, Ph.D., UG-, and PG-level students. He is associated with many research, academic, and professional societies in various capacities.

# Contributors

**Anushree Acharya**
DOMS, SGSITS
Indore, Madhya Pradesh, India

**Gaydaa AlZohbi**
Department of Mechanical Engineering, Prince Mohammad Bin Fahd University
Al Khobar, Saudi Arabia

**Rakesh Kr. Arora**
Dr. Akhilesh Das Gupta Institute of Technology and Management
Delhi, India

**Farshad Badie**
Faculty of Computer Science and Informatics, Berlin School of Business and Innovation
Berlin, Germany

**Bhawana Bhardwaj**
HPKV Business School, School of Commerce and Management Studies, Central University of Himachal Pradesh
Dharmshala, Himachal Pradesh, India

**Rohan Bhargava**
Amity School of Business, Amity University
Noida, Uttar Pradesh, India

**Meenakshi Bisla**
School of Management, The NorthCap University, Sector 23A
Gurgaon, Haryana, India

**Nitin Kumar Chauhan**
Department of Mechanical Engineering, JSS Academy of Technical Education
Noida, Uttar Pradesh, India

**Prashant Chauhan**
Department of Mechanical Engineering, JSS Academy of Technical Education
Noida, Uttar Pradesh, India

**Preeti Chawla**
School of Management, The NorthCap University
Gurgaon, Haryana, India

**Amit Dangi**
Department of Management and Commerce, The NorthCap University
Gurugram, Haryana, India

**Megha Garg**
Department of Management Studies, J.C. Bose University of Science and Technology, YMCA
Faridabad, Haryana, India

**Manoj Kr. Gupta**
Bhagwan Parshuram Institute of Technology
Delhi, India

**Neha Gupta**
Amity School of Business, Amity University
Noida, Uttar Pradesh, India

**Naveen J.**
Department of Mechanical Engineering, JSS Academy of Technical Education
Noida, Uttar Pradesh, India

**Rishav Katoch**
Department of Computer Science and Engineering, Chandigarh University
Mohali, Punjab, India

**Ajay Kumar**
Department of Mechanical Engineering, School of Engineering and Technology, JECRC University
Jaipur, Rajasthan, India

**Gulab Kumar**
HPKV Business School, School of Commerce and Management Studies, Central University of Himachal Pradesh
Dharmshala, Himachal Pradesh, India

**Parveen Kumar**
Department of Management Studies, J.C. Bose University of Science and Technology, YMCA
Faridabad, Haryana, India

**Pranshu Mishra**
Department of Computer Science and Engineering, Chandigarh University
Mohali, Punjab, India

**Chand Prakash**
Faculty of Commerce and Management, SGT University
Gurugram, Haryana, India

**Muddapu Raja**
VIT Business School, VIT University
Vellore, Tamil Nadu, India

**Hem Shweta Rathore**
Amity School of Business, Amity University
Noida, Uttar Pradesh, India

**Kunal Rawal**
Department of Hospital Administration, Sri Aurobindo Institute of Management and Science (SAIM&S)
Indore, Madhya Pradesh, India

**Nyasha Saurabh**
Department of Computer Science and Engineering, Chandigarh University
Mohali, Punjab, India

**Savita**
Amity College of Commerce, Amity University
Gurugram, Haryana, India

**K. Tara Shankar**
Indirapuram Institute of Higher Studies
Ghaziabad, Uttar Pradesh, India

**Dipanker Sharma**
HPKV Business School, School of Commerce and Management Studies, Central University of Himachal Pradesh
Dharmshala, Himachal Pradesh, India

**Renuka Sharma**
Amity College of Commerce, Amity University
Gurugram, Haryana, India

**Richa Sharma**
Department of Computer Science and Engineering, Chandigarh University
Mohali, Punjab, India

**Amardeep Singh**
School of Commerce, Manav Rachna International Institute of Research and Studies (Deemed University)
Faridabad, Haryana, India

**Suraj Pal Singh**
Chandigarh University
Mohali, Punjab, India

**Shweta Soni**
FMS, WISDOM, Banasthali Vidyapith
Rajasthan, India

**T. N. V. R. L. Swamy**
VIT Business School, VIT University
Vellore, Tamil Nadu, India

**Palanivel Rathinasabapathi Velmurugan**
Faculty of Economics and Business Administration, Berlin School of Business and Innovation
Berlin, Germany

# *Section 1*

## *Disruptive Technologies in Operations*

# 1 Embracing Disruption

## *Emerging Technologies Reshaping the Financial Services Sector*

*Renuka Sharma and Savita*

### 1.1 INTRODUCTION

High-tech breakthroughs and digital evolution indicate a significant move around in the banking industry. The cornerstone of the above-discussed technological transition is advancements in business frameworks based on immersive technology that are available to customers [17, 62]. In addition, the yardstick for developments in finance technology usefulness is based on newness and the power to offer better versatile financial services (FSs) to clients than the existing systems. Bank services and financial innovative advancements used to compete, but today they work together [16].

The blooming of financial high-tech innovations and inventions suggests the following: (i) multichannel consumer aid: personal computers, smart mobiles, PC tablets, notebook computers, or smart devices; (ii) cloud computing is another milestone that removes the need for a physical setup in terms of a proper office; (iii) cryptocurrency acceptance spreads the reach in terms of extensive reach for transactions, not many intermediaries and crystal clear channels; and (iv) digital device payments that provide safety and speed. It symbolizes very innovative methods and techniques of business and handling wealth [22, 42, 90].

Digitizing and device networking have enabled the rise of banking websites and organizations specializing in digital payment methods, online financing and social fundraising, virtual broking, and digital transfers.

The origin of the term Fintech lies in the roots of the concepts of finance and technology that refer to imminent technologies that have completely changed the picture of the financial industry [81]. Therefore, financial technology combines technological improvements with financial processes, commodities, and services [83].

In this scenario, technological innovation in finance is being utilized for attracting traditional banking clients, maximizing business transactions, improving lending and advertising, and making banking procedures easier [55, 91, 97]. As a result, Fintech is nothing else but an upgraded version of existing systems with incorporation of latest technological innovations and big data [20, 39, 60, 94].

DOI: 10.1201/9781032700953-2

In this sense, the blockchain notion is a database that is shared and serves as a platform for registering purchase and sale transactions or some other variety of transaction, ensuring the conservation of information that cannot be lost, edited, or destroyed. It is secure against manipulations and fraud since it uses cryptographic keys and is spread across several computers [18, 34, 84]. Any sort of information that must be retained intact and made available may be cache on the blockchain platform in a safe, dispersed, and cost-effective manner rather than through middlemen. Its applications range from economics to healthcare, cybersecurity, and Internet of Things (IoT). In parlance to this research, it is vital to realize that blockchain provides verification, validation, tracking, and storage of information concerning money and financial transactions [1, 67].

A blockchain is a computerized transaction ledger. The meaning lies in the word itself as it breaks the data or information into different categories and then connects all the blocks into one chain. Aside from tracking monetary transactions like Bitcoin, blockchains have a big list of probable uses. Data is saved and stored on a blockchain, making it harder to alter, hack, or cheat the network [7, 68]. A blockchain is a computer network that replicates and distributes digital transaction records. Trending technology has been in use in the banking sector to guarantee the security of data and processes. It has already found traction in the financial industry. It provides secure, trustworthy, and verifiable monetary transactions, as seen by the development of cryptocurrencies [82].

Blockchain is an online database that allows for real-time updating of certain operational information across several devices. This technique maintains digital data from transactions, contracts, and personal databases in the form of connected blocks [19]. Blockchain technology solves most of these concerns while also significantly lowering financial risk. The worldwide technology sector is beginning to recognize the advantages of blockchain technology. A particular section of eminent business community always understands its importance and at the same time explores how to leverage this technology and its benefits into their businesses.

The paramount goal of founding banks was to bring the population together and allow them to engage securely and effectively through trade and business. The blockchain-based platform is a technology that allows for the accomplishment of multiple activities on a global scale [29, 31, 38].

## 1.2 LITERATURE REVIEW

Carbó-Valverde [27, 69] asserted that a breakthrough financial technology would be challenging to govern and administer. A significant difficulty is maintaining an effective supervision over the emerging financial entities. Nevertheless, the author mentioned that Fintech might potentially be leveraged to increase efficiency and lower margins in the banking and finance business. The banking and FS industry will require new hires and new skill sets. Banking is going through internal turmoil in addition to the fact that banking and Financial Services need to rethink themselves and prepare for new financial technology. Furthermore, not only will the technology be a disruptor, but also the businesses that use it will have an impact on how it is applied, resulting in market upheaval.

Singh and Singh [78] underlined that it is a safe platform for exchanging digital currency and conducting agreements and transactions. He said that every member of the block has access to the latest edition of the coded ledger, which enables them to verify the authenticity of a new transaction. The report also underlined the significance of blockchain and how it will affect the future of banking and financial institutions, as well as how it may be used in connection with the IoT.

Brandon [14, 57] covered several blockchain uses in Financial Services. They have mostly concentrated on developing apps for the core protocol and constructing a secure and private blockchain. They further elaborated on blockchain accounting applications as "triple entry bookkeeping" as three persons are involved: the purchaser, the vendor, and the blockchain network. They further explored blockchain's possible use in financial transactions. According to the report, technology only accounts for 20% of the barriers to adoption, with the other 80% ascribed to current business processes and models.

Kaur et al. [47, 75] described blockchain as a fundamental technology with the potential to significantly lower transaction costs and transform the economy. It has the potential to impact any agreement, procedure, job, or payment. They also did a thorough analysis of the effects of blockchain technology for accounting professionals and the broader industry.

Hilary and Liu [40, 58, 87] stated that digital technology has enabled more collaboration. Cloud-based tools with tailored analytics have transformed financial accounting methods. Accounts payable and receivable, contract management, reporting, and other issues are among the cases. Cash, cashier's checks, and electronic fund transfers are the most secure payment options. Inevitably, wire transactions need time and money. Blockchain transactions eliminate these concerns, which boosts client confidence. Technology enables real-time payments between financial institutions, reducing friction and speeding up settlements. This technology can automate processes and track transactions effectively. Smart contracts can help Financial Service providers track buyer payments and seller deliverables.

## 1.3 BLOCKCHAIN TOOLS AND STRATEGIES FOR FINANCIAL SERVICES

The potential for blockchain implementation in finance has increased due to its significant benefits over conventional electronic records. A shared electronic database may be established thanks to digital ledger technology. Therefore, transaction data may be managed and stored without the need for a third party. Using blockchain technology can lessen the possibility of transaction data manipulation because there isn't a central database for storing transaction data using a special protective method [6, 32].

Some of the most important problems facing the banking sector can be resolved with the use of decentralized blockchain applications [12, 24, 89]. Across the globe, financial products and services are still provided in a traditional, centrally managed, and multilayered way. Since most financial information is kept in centrally located databases and must go via several intermediaries, it is less transparent. Furthermore, the only elements affecting the safety of information are intermediaries and the integrity of databases. However, data theft and hacking breaches may occur in even the safest databases. A lack in openness usually results in serious security difficulties since nobody is mindful of any variations until an information compromise or a different issue is discovered [4, 33, 35, 95].

Almost every business in the world might undergo a change because of the technology of blockchain. Businesses may now boost productivity, candor, and predictability in various kinds of business processes and contracts thanks to the expanding use of blockchain technology and applications.

Financial institutions are investigating ways to fully utilize blockchain technology, including offering product choices, resolving compliance issues, and removing roadblocks to risk discovery and measurement as well as associated controls [9, 53, 73].

## 1.4 VARIOUS NOTABLE BLOCKCHAIN TECHNOLOGY SERVICES IN THE FINANCIAL INDUSTRY

Blockchain-based technologies create a quick, safe, and secure record of important acts. Thus, it is ideal for transferring cash and oversees payments. On the blockchain, every step might be controlled automatically improving procedure efficiency and lowering the number of middlemen that are frequently required in these kinds of transactions. Because the blockchain platform eliminates the need for banks to settle operations, it can reduce the cost of payments [2, 41].

Financial sector bodies and blockchain experts assert that blockchain improves security, decreases risk, and saves money by boosting understanding and decreasing hassle along the lengthy chain of operations that often precedes finance encounters. To some extent, financial organizations may save money by utilizing these blockchain features. Historically, financial institutions have acted as time-consuming, complicated intermediaries amid several parties, hampering fast transactions [44, 93].

Blockchain technology instantly and securely establishes a foolproof record of crucial activities. This makes it fantastic for payments abroad and transfer of money. By automating all the steps on the blockchain, effectiveness in operation may be boosted and the dependence on middlemen that are frequently required in these kinds of trades can be lowered. By doing away with the requirement for banks to clear dealings, blockchain-based technology may lower the overall expense [2].

More market-oriented governance arrangements are made possible by electronic agreements, which may lower charges related to knowledge gather and the process, contract and negotiation drafting, agreement monitoring and implementation, and management of relationships. Because of its robust repository and guarantee that activities are carried out automatically with no need for input from humans or the participation of mediators in the payment process, electronic contracts can boost data confidence [48, 64, 76].

## 1.5 APPLICATIONS OF BLOCKCHAIN TECHNOLOGY IN THE FINANCE AREA

| S.No. | a. Usage | b. Description |
|---|---|---|
| 1. | Fraud-prevention area | High levels of security could be achieved when decentralization and third-party elimination assuming the invariable traits of blockchain technology systems and innovative agreements (smart contracts). The proposed blockchain, which incorporates fraud-detection technology, is expected to help reduce fraudulent credit card transactions that involve intermediaries [49]. |
| 2. | Banking institutions | Both public and private blockchains may be used in a wide range of financial use cases, creating fresh markets for banking offerings that are favourable to bankers and clientele alike by enabling quicker, less expensive, more reliable, and far more comprehensive trades.<br>In addition, applications based on blockchain promote the growth of multi-polar methods, which will boost the productivity of the FS sector.<br>Blockchain facilitates direct one-to-one transaction within the financial sector.<br>The successful management of transactions by smart contracts will result in the removal of intermediaries. As the number of system layers is reduced, instantaneous payment settlements become feasible. Moreover, blockchain payment systems enable instantaneous cross-border payments [15, 36, 70]. |
| 3. | Privacy and security | Blockchain's personal and hybrid networks are strong, and it is being built in such a way so that it can manage numerous transactions per second and abrupt spikes in network usage.<br>Payment delay is there, sometimes in the banking sector as payments may get stuck for almost a week. Due to the involvement of various intermediaries in the financial structure, completion of a transaction is a little lengthy process.<br>Blockchain technology is an effective method of ensuring quick transactions without compromising confidentiality and privacy. Using technologies like smart contracts and digital ledgers, more efficiency can be brought into banking [37, 66, 100]. |
| 4. | Secure domestic and overseas payments | Blockchain science makes straightforward and secure nationwide and worldwide payments feasible with little-to-no mediators. Additionally, blockchain can streamline the process and significantly reduce costs. Users can manage and keep their data on a blockchain by using blockchain technology. According to the industry, blockchain is attractive because it can lead to cost savings, new levels of performance of operations, and improvements in accuracy and transparency.<br>The decentralization and one-to-one trading enabled by distributed ledger technology and blockchain are contributing to the global increase in the efficiency of financial transactions. These technologies, by their very nature, have proven to be highly effective in expediting traditional transactional procedures and providing quick worldwide payment options [21, 43, 45]. |

*(Continued)*

| S.No. | a. Usage | b. Description |
|---|---|---|
| 5. | Sharing of knowledge and collaboration | Collective knowledge verification and integration are possible with the inclusion of blockchain science [56].<br>Any knowledge that is produced can be stored in a blockchain database, made completely transparent to the public, and utilized by users. Following the initial extraction, the data can be time-stamped and added to a blockchain. The encrypted data is made public after a predetermined amount of time spent hidden.<br>Applications for blockchain technology are numerous in the disciplines of financial reporting, health coverage, and trade finance, payments, settlement processes, and auditing purposes [8, 28, 63, 65]. |
| 6. | Boosts digital payments | Digital payment and mobile banking transactions have grown quickly since the launch of UPI and Fintech. Because of its increased integration with other economic endeavours, identification and authentication have become problematic. Here, blockchain technology aids in guaranteeing users' digital identities [61]. |
| 7. | Identity management | Identification handling is yet another fantastic area in which blockchain-based technology is being utilized.<br>Blockchain technology will be helpful to users; they can have their own digital identity, which is comparatively reliable, safe, and resistant. In near future, blockchain-based technologies will be replacing the concept of usernames and passwords. The concept of usernames is more sensitive to privacy issues.<br>It will become easy for people to sign digital docs and to login into various web pages and platforms. Blockchain technology has the power to substitute the business tasks that are more based on transactions [92, 99]. |
| 8. | Financial services | The practical use of the blockchain system has increased in the invention and innovation of new technologies of financial areas. Over the years, the investments in the newer and newer financial applications have increased dramatically. Blockchain adoption by numerous businesses throughout the financial system is happening a little bit. Blockchain technology is a great tool that is compatible with millions of endpoints and large databases [3, 23, 71]. |
| 9. | Tokenization | Blockchain technology also helps in the process of tokenization. Under this process, on a blockchain, tickets are made to represent tangible resources. Central Bank Digital Currency also contributed to the need for blockchain.<br>Additionally, a simple token system could increase a platform's user base and attract banking services. The blockchain is mostly utilized for artificial intelligence (AI)–based decision-making and storing client requests. The payment business has been gradually impacted by blockchain technology, which has changed the nature of transactions. Implementing new financial procedures and infrastructure, eliminated corruption and promoted efficiency and simplicity. The world of banking is being transformed by the technology known as blockchain [30, 75]. |

*(Continued)*

| S.No. | a. Usage | b. Description |
|---|---|---|
| 10. | Smart contracts | Smart contractual arrangements are essential for streamlining processes and settlements because they provide instantaneous and definite results for all parties involved without the need for mediators or obstacles. A combination of AI and smart contracts can bring dynamic changes in banking efficiency. The use of smart contracts for loan regulation will boost the implementation and services of the loan departments [77, 80, 96]. |
| 11. | Boosting productivity | The advancements in the area of the block technology have resulted into tremendous positive changes in the area of settlement systems of transactions. By doing away with middlemen, this technology makes global trade of transactions easy, more secure, and affordable. As blockchain adoption grows, this will probably cause a significant change in the international system of clearance (transaction settlement) outlook, enabling transactions easy at global level [5, 25, 46, 72]. |
| 12. | Digital asset management | The asset management industry has a lot of opportunities. On the one hand, the distributed and unalterable nature of blockchain makes it possible to track and transfer assets with greater transparency and security. On the other hand, Supply Chain Technology, which is an essential aspect of Blockchain Technology, simplifies and automates various processes, cutting operational costs and limiting the risk of errors. Furthermore, this technology makes it easier to settle real-time disputes and improve liquidity, which further increases the efficiency of asset management workflows.<br>Careful thought must also be given to security issues pertaining to the storage and safeguarding of digital assets as well as the capacity scaling of blockchain networks. Moreover, a paradigm shift in corporate practices and organizational structures is required for the inclusion of blockchain in asset management. Stakeholders must embrace the technological and cultural shifts brought about by the implementation of BT systems and adjust to new methods of asset management and transfer [10, 13, 88]. |
| 13. | Improve client affordability | Within the financial industry, blockchain takes the responsibility of removing the menace of deceit, boosting transparency and increasing affordability of products for clients. It could significantly boost transparency within the finance sector. The use of blockchain can ensure fraud detection and inefficiency removal, which ultimately ensures less risk in the FS industry. Customers are more active virtually in today's world because more businesses are available online. This serves as an open invitation to scammers. It might allay this concern [90, 98].<br>Compared to traditional banking, block-based clearances and transfers are swifter and trackable [54, 55, 56, 86]. |
| 14. | Data registries and repositories | Numerous data registries and repositories that offer the verified data required for a range of transactions are essential to modern finance.<br>While these data repositories are crucial, there are certain challenges associated with it like duplicity issues, vulnerable privacy, and excess cost. The blockchain can be the best answer for the infrastructural problems in the financial sector [13, 59, 85, 101]. |

*(Continued)*

| S.No. | a. Usage | b. Description |
|---|---|---|
| 15. | Customized investment management | Tiny investors have always been able to involve in the asset markets through mutual funds, provided they are ready to accept a defined menu of products. This market is good for investors who want to track a well-known benchmark index; however, investors seeking non-linear payoffs or exposure to customized indexes should look elsewhere [11, 62, 69]. |
| 16. | Insurance claim management | As use cases, blockchains provide some very attractive benefits for the insurance industry. They are a great way to speed up the laborious process of managing insurance claims, to give one notable example. Processing claims is indeed becoming a highly drawn-out and complex procedure that involves multiple parties and intermediaries before a claim request is deemed final and payment can be made. Based on the execution of smart contracts to realize all checks and verifications and the integration of all stakeholders around a distributed ledger infrastructure, blockchains can facilitate this procedure. A smart contract can compute and validate the amount that needs to be paid, among all other necessary stages in the claim processing process, automatically and securely [26, 38, 50, 51, 52]. |

## 1.6 CHALLENGES ASSOCIATED WITH BLOCKCHAIN

Despite all its potential, several challenges could keep blockchain from being broadly used. Blockchain is a dispensed one-to-one technology that enables any network user to access transaction databases and insert new data to the databases.

The platform's underlying principles of increased openness and limited centralized oversight have negative consequences and hinder the effective application of Blockchain Technology.

Some issues are still present in the financial markets that need to be fixed, i.e., latency, security, privacy, and scalability.

### 1.6.1 Issue of Scalability

Ethereum platform is used in blockchain. At a time, only 20 entries can be handled by Ethereum and Bitcoin. It indicates that it is not possible to meet the demand of handling millions of data files quickly. The cause is that blocks have a limited capacity [13].

### 1.6.2 Security

Systems designed on blockchain technology are susceptible. Blockchain's security remains compromised, as evidenced by numerous past attacks, attributed to self-serving miners.

### 1.6.3 Privacy Issue

To avert information leaking is one of the key objectives. Blockchain can create multiple addresses for users instead of using their actual identification, which is believed to enhance security. However, because all transactions and balance data are publicly accessible, blockchain cannot put a stop to the leakage of transactional information.

### 1.6.4 Energy Consumption

The operational and storage expenses of big data programs might surpass the long-term storage costs associated with transaction data and electronic fund transfers. [74] pointed out that the computational demands for blockchain operations are escalating rapidly. The network of Bitcoins consumes a substantial percentage of electricity, with just one Bitcoin deal [79, 85].

### 1.6.5 Expensive Technology

Transitioning to this new technology can be a costly and effort-intensive endeavour, due to the shortage of skilled professionals. It will be a challenge for small financial institutions because of the heavy cost of updating their current systems.

## 1.7 CONCLUSION

Globally, factories are adopting blockchain technology as they become increasingly networked. Creating an impenetrable ledger for digital assets like cryptocurrency. Security of ledgers is a crucial objective of block-based systems. The absorption of this disruptive concept has increased manyfold in the banking system. Since bank accounts are the primary means of exchanging money, settlements are indispensable. With the issuance of their own virtual currencies and use of innovative technologies to guarantee trustworthy settlements, banks are in the central role for the digital shift. Bankers are now able to keep an eye on every trade in real time because of digital ledger technology. These agreements allow businesses employing blockchain-based platforms in finance to post invoices on the network of ledgers. Dates, quantities, and details about the patron related to payments can all be found on blockchain. The status of payments by the consumer can be immediately updated to the businesses. Before transacting, blockchain computing in finance has the potential to assess the client's trustworthiness. A dispensed ledger system, such as the BT, is encrypted by two types of security keys. As it operates on a decentralized ledger, blockchain will manage several operations in the finance industry. The financial world is well informed about the futuristic rewards of blockchain science and that, going forward, it will have a substantial impact on FSs. Blockchain technology has contributed tremendous scientific contributions, propelling innovation across several sectors. As technological advances continue, the potential for transforming industry and society remains high. It has a significant contribution viz. data security and privacy, data integrity and transparency, decentralized finance. Future study should focus on the banking sector, FS trade, territorial growth, legal and managerial issues, research methodology, and financial technology sustainability. It was confirmed that there is an increasing and progressive interest in research effort on financial technology on a global scale.

## REFERENCES


1. Abad-Segura, Emilio, and Mariana-Daniela González-Zamar. "Global research trends in financial transactions." *Mathematics* 8, no. 4 (2020): 614.
2. Alabbasi, Yousef. "Governance and legal framework of blockchain technology as a digital economic finance." *International Journal of Innovation in the Digital Economy (IJIDE)* 11, no. 4 (2020): 52–62.
3. Alao, Olakunle, and Paul Cuffe. "Towards a blockchain special purpose vehicle for financing independent renewable electricity projects in sub-saharan africa." In *2020 6th IEEE International Energy Conference (ENERGYCon)*, pp. 1041–1046, IEEE, 2020.
4. Alao, Olakunle, and Paul Cuffe. "Structuring special purpose vehicles for financing renewable generators on a blockchain marketplace." *IEEE Transactions on Industry Applications* 58, no. 2 (2021): 1478–1489.
5. An, Yun Joo, Paul Moon Sub Choi, and Seth H. Huang. "Blockchain, cryptocurrency, and artificial intelligence in finance." In S. H. Huang & P. M. S. Choi (Eds.), *Fintech with Artificial Intelligence, Big Data, and Blockchain* (pp. 1–34). Singapore: Springer Verlag, Singapore, 2021.
6. Ankenbrand, Thomas. "A structure for evaluating the potential of blockchain use cases in finance." *Perspectives of Innovations, Economics & Business* 17, no. 2 (2017): 77–95.
7. Anoop, V. S., and Justin Goldston. "Decentralized finance to hybrid finance through blockchain: a case-study of acala and current." *Journal of Banking and Financial Technology* 6, no. 1 (2022): 109–115.
8. Ariffin, Nizamuddin, and Ahmad Zuhairi Ismail. "The design and implementation of trade finance application based on hyperledger fabric permissioned blockchain platform." In *2019 International Seminar on Research of Information Technology and Intelligent Systems (ISRITI)*, pp. 488–493, IEEE, 2019.
9. Aysan, Ahmet Faruk, Behar Sadriu, and Humeyra Topuz. "Blockchain futures in cryptocurrencies, trade and finance: A preliminary assessment." *Bulletin of Monetary Economics and Banking* 23, no. 4 (2020): 525–542.
10. Babich, Volodymyr, and Gilles Hilary. "Blockchain and other distributed ledger technologies in operations." *Foundations and Trends® in Technology, Information and Operations Management* 12, no. 2–3 (2019): 152–172.
11. Balagolla, E. M. S. W., W. P. C. Fernando, R. M. N. S. Rathnayake, M. J. M. R. P. Wijesekera, A. N. Senarathne, and K. Y. Abeywardhana. "Credit card fraud prevention using blockchain." In *2021 6th International Conference for Convergence in Technology (I2CT)*, pp. 1–8, IEEE, 2021.
12. Bhogal, Tarsem, and Arun Trivedi. *International Trade Finance, A Pragmatic Approach* (pp. 303–312). Cham: Springer.
13. Biais, Bruno, Christophe Bisiere, Matthieu Bouvard, and Catherine Casamatta. "The blockchain folk theorem." *The Review of Financial Studies* 32, no. 5 (2019): 1662–1715.
14. Brandon, Dan. "The blockchain: The future of business information systems." *International Journal of the Academic Business World* 10, no. 2 (2016): 33–40.
15. Buitenhek, Mark. "Understanding and applying blockchain technology in banking: Evolution or revolution?" *Journal of Digital Banking* 1, no. 2 (2016): 111–119.
16. Basole, Rahul C., and Shiv S. Patel. "Transformation through unbundling: Visualizing the global FinTech ecosystem." *Service Science* 10, no. 4 (2018): 379–396.
17. Breidbach, Christoph F., Byron W. Keating, and Chiehyeon Lim. "Fintech: Research directions to explore the digital transformation of financial service systems." *Journal of Service Theory and Practice* 30, no. 1 (2020): 79–102.
18. Cahill, Daniel, Dirk G. Baur, Zhangxin Frank Liu, and Joey W. Yang. "I am a blockchain too: How does the market respond to companies' interest in blockchain?" *Journal of Banking & Finance* 113 (2020): 105740.

19. Caldarelli, Giulio, and Joshua Ellul. "The blockchain oracle problem in decentralized finance—a multivocal approach." *Applied Sciences* 11, no. 16 (2021): 7572.
20. Chang, Hsiu-yun, Woan-lih Liang, and Yanzhi Wang. "Do institutional investors still encourage patent-based innovation after the tech bubble period?" *Journal of Empirical Finance* 51 (2019): 149–164.
21. Chen, Jingjing, Shiyi Chen, Qingfu Liu, and M. I. Shen. "Applying blockchain technology to reshape the service models of supply chain finance for SMEs in China." *The Singapore Economic Review* 66, no. 3 (2021): 1–18.
22. Chen, Mark A., Qinxi Wu, and Baozhong Yang. "How valuable is FinTech innovation?" *The Review of Financial Studies* 32, no. 5 (2019): 2062–2106.
23. Chen, Yi-Hui, Li-Chin Huang, Iuon-Chang Lin, and Min-Shiang Hwang. "Research on the secure financial surveillance blockchain systems." *International Journal of Network Security* 22, no. 4 (2020): 708–716.
24. Chen, Tung-Chun, Yu-Shen Liang, Po-Sheng Ko, and Jui-Chan Huang. "Optimization model of cross-border e-commerce payment security by blockchain finance." *Wireless Communications and Mobile Computing* 2021, no. 1 (2021): 9192219.
25. Cohen, Lewis Rinaudo, Lee Samuelson, and Hali Katz. "How securitization can benefit from blockchain technology." *Journal of Structured Finance* 23, no. 2 (2017): 51.
26. Collomb, Alexis, and Klara Sok. "Blockchain/distributed ledger technology (DLT): What impact on the financial sector?" *Digiworld Economic Journal* 103, no. 3 (2016): 93.
27. Carbó-Valverde, Santiago. "The impact on digitalization on banking and financial stability." *Journal of Financial Management, Markets and Institutions* 1 (2017): 133–140.
28. Dang, Chenlu, Fan Wang, Zimo Yang, Hongxia Zhang, and Yufeng Qian. "Retraction Note: Evaluating and forecasting the risks of small to medium-sized enterprises in the supply chain finance market using blockchain technology and deep learning model." *Operations Management Research* 17, no. 2 (2024): 1.
29. Du, Mingxiao, Qijun Chen, Jie Xiao, Houhao Yang, and Xiaofeng Ma. "Supply chain finance innovation using blockchain." *IEEE transactions on Engineering Management* 67, no. 4 (2020): 1045–1058.
30. Elrefae, Ghaleb, and M. Nuseir. "Blockchain in global finance make-over: Exploring the mediating role of supply chain flexibility." *Uncertain Supply Chain Management* 10, no. 3 (2022): 983–992.
31. Eyal, Ittay. "Blockchain technology: Transforming libertarian cryptocurrency dreams to finance and banking realities." *Computer* 50, no. 9 (2017): 38–49.
32. Gao, Wenyou, and Chang Su. "Analysis of earnings forecast of blockchain financial products based on particle swarm optimization." *Journal of Computational and Applied Mathematics* 372 (2020): 112724.
33. Gorkhali, Anjee, and Rajib Chowdhury. "Blockchain and the evolving financial market: A literature review." *Journal of Industrial Integration and Management* 7, no. 1 (2022): 47–81.
34. Guégan, Dominique, and Christophe Hénot. "A probative value for authentication use case blockchain." *Digital Finance* 1, no. 1 (2019): 91–115.
35. Guo, Lingling, Jingjing Chen, Shihan Li, Yafei Li, and Jinzhi Lu. "A blockchain and IoT-based lightweight framework for enabling information transparency in supply chain finance." *Digital Communications and Networks* 8, no. 4 (2022): 576–587.
36. Guo, Ye, and Chen Liang. "Blockchain application and outlook in the banking industry." *Financial Innovation* 2 (2016): 1–12.
37. Haleem, Abid, Mohd Javaid, Ravi Pratap Singh, Rajiv Suman, and Shanay Rab. "Blockchain technology applications in healthcare: An overview." *International Journal of Intelligent Networks* 2 (2021): 130–139.

38. Harwick, Cameron, and James Caton. "What's holding back blockchain finance? On the possibility of decentralized autonomous finance." *The Quarterly Review of Economics and Finance* 84 (2022): 420–429.
39. Hendrikse, René. "Can selfies spark the identity (r) evolution in financial services?" *Biometric Technology Today* 2019, no. 4 (2019): 5–7.
40. Hilary, Gilles, and Laura Xiaolei Liu. "Blockchain and other distributed ledger technologies in finance." *The Palgrave Handbook of Technological Finance* 1 (2021): 243–268.
41. Iman, Nofie. "Assessing the dynamics of Fintech in Indonesia." *Investment Management and Financial Innovations* 15, no. 4 (2018): 296–303.
42. Ioannou, Ilias, and Guven Demirel. "Blockchain and supply chain finance: A critical literature review at the intersection of operations, finance and law." *Journal of Banking and Financial Technology* 6, no. 1 (2022): 83–107.
43. Jain, Neha, and R. R. Sedamkar. "A blockchain technology approach for the security and trust in trade finance." In *2020 14th International Conference on Innovations in Information Technology (IIT)*, pp. 192–197, IEEE, 2020.
44. Jiang, C., and C. Ru. "Application of blockchain technology in supply chain finance." In *2020 5th International Conference on Mechanical, Control and Computer Engineering (ICMCCE),* pp. 1342–1345, IEEE, 2020, December.
45. Karagiannis, I., K. Mavrogiannis, J. Soldatos, D. Drakoulis, E. Troiano, and A. Polyviou. "Blockchain based sharing of security information for critical infrastructures of the finance sector." In A. P. Fournaris, K. Lampropoulos, & E. M. Tordera (Eds.), *International Workshop on Information and Operational Technology Security Systems* (pp. 226–241). Cham: Springer International Publishing, 2019, September.
46. Karajovic, Maria, Henry M. Kim, and Marek Laskowski. "Thinking outside the block: Projected phases of blockchain integration in the accounting industry." *Australian Accounting Review* 29, no. 2 (2019): 319–330.
47. Kaur, Jaspreet, Satish Kumar, Balkrishna E. Narkhede, Marina Dabić, Ajay Pal Singh Rathore, and Rohit Joshi. "Barriers to blockchain adoption for supply chain finance: The case of Indian SMEs." *Electronic Commerce Research* 24, no. 1 (2024): 303–340.
48. Kherbouche, Meriem, Galena Pisoni, and Bálint Molnár. "Model to program and blockchain approaches for business processes and workflows in finance." *Applied System Innovation* 5, no. 1 (2022): 10.
49. Kim, Henry, Ushnish Sengupta, and Marek Laskowski. "Business in the front, crypto in the back: How the blockchain startup, novera, is innovating upon traditional finance." *International Review of Entrepreneurship* 19, no. 1 (2021): 117–136
50. Kosba, Ahmed, Andrew Miller, Elaine Shi, Zikai Wen, and Charalampos Papamanthou. "Hawk: The blockchain model of cryptography and privacy-preserving smart contracts." In *2016 IEEE Symposium on Security and Privacy (SP)*, pp. 839–858, IEEE, 2016.
51. Kube, Nicolas. "Daniel Drescher: Blockchain basics: A non-technical introduction in 25 steps: *Apress* 32 (2017): 329–331.
52. Kucukaltan, Berk, Rifat Kamasak, Baris Yalcinkaya, and Zahir Irani. "Investigating the themes in supply chain finance: the emergence of blockchain as a disruptive technology." *International Journal of Production Research*, 62, no. 22 (2022): 1–20.
53. Lee, Jong-Hyouk, and Marc Pilkington. "How the blockchain revolution will reshape the consumer electronics industry [future directions]." *IEEE Consumer Electronics Magazine* 6, no. 3 (2017): 19–23.
54. Li, Yunhe, Yu Liu, and Feixue Xie. "Technology directors and firm innovation." *Journal of Multinational Financial Management* 50 (2019): 76–88.

55. Li, Zhi, Ray Y. Zhong, Zong-Gui Tian, Hong-Ning Dai, Ali Vatankhah Barenji, and George Q. Huang. "Industrial blockchain: A state-of-the-art Survey." *Robotics and Computer-Integrated Manufacturing* 70 (2021): 102124.
56. Liebenau, Jonathan, and Silvia Elaluf-Calderwood. "Blockchain innovation beyond bitcoin and banking." *Available at SSRN* 2749890 (2016): 1–16.
57. Maurer, Bill. "Re-risking in realtime. On possible futures for finance after the blockchain." *Behemoth-A Journal on Civilisation* 9, no. 2 (2016): 82–96.
58. Meiklejohn, Sarah, Marjori Pomarole, Grant Jordan, Kirill Levchenko, Damon McCoy, Geoffrey M. Voelker, and Stefan Savage. "A fistful of bitcoins: Characterizing payments among men with no names." In *Proceedings of the 2013 Conference on Internet Measurement Conference*, pp. 127–140, 2013.
59. Mention, Anne-Laure. "The future of fintech." *Research-Technology Management* 62, no. 4 (2019): 59–63.
60. Mhlanga, David. "Block chain technology for digital financial inclusion in the industry 4.0, towards sustainable development?" *Frontiers in Blockchain* 6 (2023): 1035405.
61. Müller, János, and Ádám Kerényi. "The need for trust and ethics in the digital age–Sunshine and shadows in the FinTech world." *Financial and Economic Review* 18, no. 4 (2019): 5–34.
62. Natalia Maslova, C. M. A., and PMP CTP. "Blockchain: Disruption and opportunity." *Strategic Finance* 100, no. 1 (2018): 24–29.
63. Natanelov, Valeri, Shoufeng Cao, Marcus Foth, and Uwe Dulleck. "Blockchain smart contracts for supply chain finance: Mapping the innovation potential in Australia-China beef supply chains." *Journal of Industrial Information Integration* 30 (2022): 100389.
64. Nguyen, Quoc Khanh. "Blockchain-a financial technology for future sustainable development." In *2016 3rd International Conference on Green Technology and Sustainable Development (GTSD)*, pp. 51–54, IEEE, 2016.
65. O'Dair, Marcus, and Robyn Owen. "Monetizing new music ventures through blockchain: Four possible futures?" *The International Journal of Entrepreneurship and Innovation* 20, no. 4 (2019): 263–276.
66. O'Neill, Michael, and Anthony Brabazon. "Business analytics capability, organisational value and competitive advantage." *Journal of Business Analytics* 2, no. 2 (2019): 160–173.
67. O'Dair, Marcus, and Robyn Owen. "Financing new creative enterprise through blockchain technology: Opportunities and policy implications." *Strategic Change* 28, no. 1 (2019): 9–17.
68. Omarini, Anna. "The digital transformation in banking and the role of FinTechs in the new financial intermediation scenario." *International Journal of Finance, Economics and Trade* 1, no. 1 (2017): 1–6.
69. Owen, Robyn, Ciaran Mac an Bhaird, Javed Hussain, and Tiago Botelho. "Blockchain and other innovations in entrepreneurial finance: Implications for future policy." *Strategic Change* 28, no. 1 (2019): 5–8.
70. Pal, Abhinav, Chandan Kumar Tiwari, and Aastha Behl. "Blockchain technology in financial services: A comprehensive review of the literature." *Journal of Global Operations and Strategic Sourcing* 14, no. 1 (2021): 61–80.
71. Pan, Wenxuan, and Meikang Qiu. "Application of blockchain in asset-backed securitization." In *2020 IEEE 6th Intl Conference on Big Data Security on Cloud (BigDataSecurity), IEEE Intl Conference on High Performance and Smart Computing,(HPSC) and IEEE Intl Conference on Intelligent Data and Security (IDS)*, pp. 71–76, IEEE, 2020.
72. Patel, Ritesh, Milena Migliavacca, and Marco E. Oriani. "Blockchain in banking and finance: A bibliometric review." *Research in International Business and Finance* 62 (2022): 101718.

73. Price, D. "Big blockchain issues: Security, privacy, legal, regulatory, and ethical." (2018).
74. Saini, C. P. "Difficulty in transition: Study on hindrances in online shopping for potential shoppers." *International Journal of Online Marketing (IJOM)* 10, no. 2 (2020): 17–29.
75. Sanka, Abdurrashid Ibrahim, Muhammad Irfan, Ian Huang, and Ray C. C. Cheung. "A survey of breakthrough in blockchain technology: Adoptions, applications, challenges and future research." *Computer communications* 169 (2021): 179–201.
76. Scott, Brett. *How Can Cryptocurrency and Blockchain Technology Play a Role in Building Social and Solidarity Finance?* No. 2016-1. UNRISD Working Paper, 2016.
77. Sharma, Pradip Kumar, Neeraj Kumar, and Jong Hyuk Park. "Blockchain technology toward green IoT: Opportunities and challenges." *IEEE Network* 34, no. 4 (2020): 263–269.
78. Singh, Sachchidanand, and Nirmala Singh. "Blockchain: Future of financial and cyber security." In *2016 2nd international Conference on Contemporary Computing and Informatics (IC3I)*, pp. 463–467, IEEE, 2016.
79. Sivaram, M., E. Laxmi Lydia, Irina V. Pustokhina, Denis Alexandrovich Pustokhin, Mohamed Elhoseny, Gyanendra Prasad Joshi, and K. Shankar. "An optimal least square support vector machine based earnings prediction of blockchain financial products." *IEEE Access* 8 (2020): 120321–120330.
80. Smith, Kane J., and Gurpreet Dhillon. "Assessing blockchain potential for improving the cybersecurity of financial transactions." *Managerial Finance* 46, no. 6 (2020): 833–848.
81. Song, KeyongSeog. "Investigation of business model on fintech payment system." *The E-Business Studies* 16, no. 6 (2015): 65–94.
82. Tapscott, Alex, and Don Tapscott. "How blockchain is changing finance." *Harvard Business Review* 1, no. 9 (2017): 2–5.
83. Tian, Xinhui, Rui Han, Lei Wang, Gang Lu, and Jianfeng Zhan. "Latency critical big data computing in finance." *The Journal of Finance and Data Science* 1, no. 1 (2015): 33–41.
84. Van Eecke, Patrick, and Anne-Gabrielle Haie. "Blockchain and the gdpr: The eu blockchain observatory report." *European Data Protection Law Review* 4, no. 4 (2018): 531–534.
85. Varma, Jayanth Rama. "Blockchain in finance." *Vikalpa* 44, no. 1 (2019): 1–11.
86. Vinciguerra, Rosa, Francesca Cappellieri, and Michele Pizzo. "A disruptive innovation of blockchain in finance and banking: A jump into the future." *Intellectual Capital, Smart Technologies and Digitalization: Emerging Issues and Opportunities* 2 (2021): 159–170.
87. Wang, RenLan, and Yanhong Wu. "Application of blockchain technology in supply Chain finance of Beibu Gulf region." *Mathematical Problems in Engineering* 2021, no. 1 (2021): 5556424, 1860–1864.
88. Wang, Ziyi, Chen Bai, Zhuolun He, Guangliang Zhang, Qiang Xu, Tsung-Yi Ho, Bei Yu, and Yu Huang. "Functionality matters in netlist representation learning." In *Proceedings of the 59th ACM/IEEE Design Automation Conference*, pp. 61–66, 2022.
89. Wiatt, Reina G. "From the mainframe to the blockchain." *Strategic Finance* 100, no. 7 (2019): 26–35.
90. Wonglimpiyarat, Jarunee. "Analysis of Fintech in the banking industry." *International Journal of Business Innovation and Research* 19, no. 1 (2019): 125–138.
91. Wonglimpiyarat, Jarunee. "What is it about strategic implications of using financial models in the process of technology management?" *The Journal of High Technology Management Research* 30, no. 1 (2019): 82–90.

92. Woodside, Joseph M., Fred K. Augustine Jr, and Will Giberson. "Blockchain technology adoption status and strategies." *Journal of International Technology and Information Management* 26, no. 2 (2017): 65–93.
93. Xu, Minyi, Shujian Ma, and Gang Wang. "Differential game model of information sharing among supply chain finance based on blockchain technology." *Sustainability* 14, no. 12 (2022): 7139.
94. Yacoub, Ghassan. "Collaborative innovation and appropriability in start-ups: Evidence from the FinTech sector." In *Academy of Management Proceedings*, vol. 2017, no. 1, p. 13674. Briarcliff Manor, NY 10510: Academy of Management, 2017.
95. Younus, Dr Ahmed Muayad, and Mohanad Abumandil. "Role of smart contract technology blockchain services in finance and banking systems: Concept and core values." *Available at SSRN* 4078566 (2022): 1–13.
96. Yuyan, Li, and Wang Lan. "The role of blockchain technology on supply chain finance." In *Proceedings of the 3rd International Conference on Data Science and Information Technology*, pp. 153–158, 2020.
97. Zavolokina, Liudmila, Mateusz Dolata, and Gerhard Schwabe. "The FinTech phenomenon: Antecedents of financial innovation perceived by the popular press." *Financial Innovation* 2 (2016): 1–16.
98. Zhang, Li, Yongping Xie, Yang Zheng, Wei Xue, Xianrong Zheng, and Xiaobo Xu. "The challenges and countermeasures of blockchain in finance and economics." *Systems Research and Behavioral Science* 37, no. 4 (2020): 691–698.
99. Zhang, Shurui, Ji'an Tang, Qiao Hu, and Feng Liu. "Research on blockchain financial derivatives cluster." In *Proceedings of the 2020 3rd International Conference on Blockchain Technology and Applications*, pp. 40–46, 2020.
100. Zhang, Xiaheng, and Ying Han. "Research on the energy distribution and tendency of blockchain in the E-commerce field." *International Journal of Frontiers in Engineering Technology* 1, no. 1 (2019): 88–97.
101. Zheng, Zibin, Shaoan Xie, Ning Dai, Hong, Xiangping Chen, and Huaimin Wang. "Blockchain challenges and opportunities: A survey." *International Journal of Web and Grid Services* 14, no. 4 (2018): 352–375.

# 2 Disruptive Technologies for Energy Efficiency in Industries

*Gaydaa AlZohbi*

## 2.1 INTRODUCTION

Energy is very important in our daily life as it powers our houses, economics, and lifecycles. However, the source of energy generation and the way of using it negatively affect the global energy ecosystem. Currently, the global growth of population, urbanization, industrialization, technological advancement, and wealth led to the rise of demand for energy. According to International Energy Agency (IEA) [1], an increase of 25% of the universal primary energy consumption is predicted by 2030. This rise of energy consumption procures to increase the global growth rate of greenhouse gas emissions, as well as the global concerns about the environmental pollution and climate change. Additionally, a steady increase of energy costs has been recorded starting from the beginning of the 21st century. The change of energy costs, either rise or decrease, is considered as a challenge for companies to deal with. Universal actions have been taken, mainly by governments and businesses, to produce and use the energy in an efficient way. The energy sector is experiencing a transformative transition, influenced by the pressing requirement for climate urgency and the rising need for clean energy solutions. Energy efficiency, known also as the fifth fuel, is crucial to meet the global growth of energy demand, and minimize the greenhouse gas emissions. Moreover, energy efficiency is one of the approaches used to cope with the price volatility, even in the case of low energy cost.

Energy is an important element of the operational costs in industries. Some sectors, like chemical, metal, mining, and cement, spent around one-third of their operating budget on energy. The share of energy cost from the total operating costs for different industrial sectors is presented in Figure 2.1. The power sector accounts the highest energy costs, with a percentage of 50–80%, while the oil-refining sector accounts the lowest energy cost, of 7%, of the total operating costs. A decrease of energy consumption by 10–20% could be achieved through operational enhancements. The development of energy efficiency technologies could minimize the energy consumption by 50%, at least.

Disruptive technologies present the technology that aims to improve the way of producing, efficiency quality, and new business opportunities. They are considered as plenty of technologies used to enhance the life and they are also considered as an


DOI: 10.1201/9781032700953-3

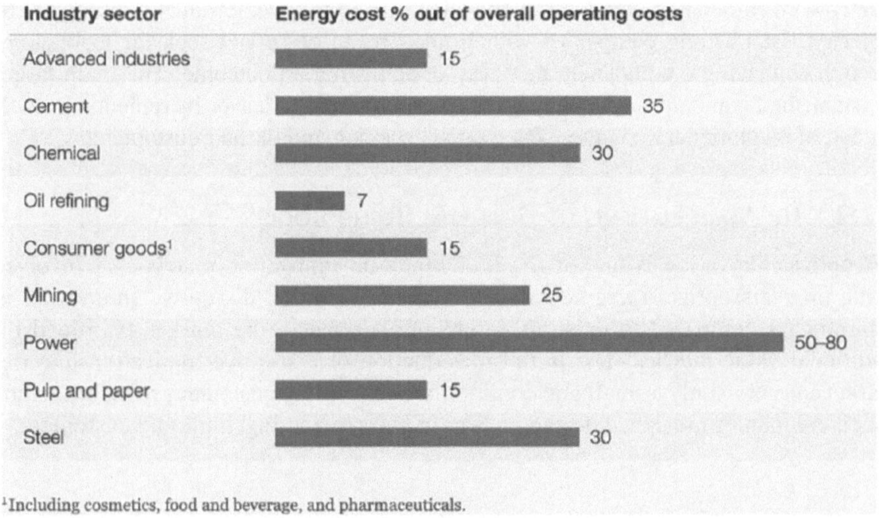


**FIGURE 2.1** Share of energy cost from the total operating costs for different industrial sectors [2].

essential component of all industrial sectors. Disruptive technologies mainly cope with the web applications, which enhanced various business opportunities by the internet as well as rising the need of internet. Customers can benefit from disruptive technologies by buying and comparing choices of various companies by being at home. Having a look at real-life examples of disruptive technologies that have converted industries over the last decades is the best way to understand disruptive innovation. The current chapter presents an overview on disruptive technologies, by presenting the history.

## 2.2 OVERVIEW ON DISRUPTIVE TECHNOLOGIES

### 2.2.1 History of Disruptive Technologies

The idea of disruptive technologies was first introduced by Clayton Christensen and Joseph L. Bower, in 1955 in *Harvard Business Review* article 'Disruptive technologies: Catching the wave'. Then, it expended on the topic in *The Innovator's Dilemma*, published by Christensen in 1997, and in the *Wall Street Journal*, *Business Week*, and *New York Times* business. A generation of businesses and entrepreneurs were inspired by the idea of disruptive technology. In 2011, the idea that companies could accurately fail since they do everything correctly was introduced corporate America owing to the revolutionary business book. In this book, Christensen clarified the reason of repulsion of successful companies by disruptive technologies since they flop to forsake traditional business practices and obsolete business models, and thus they have two options, either innovating or dying. In 2013, Christensen and Michael Raynor extended on its ideas and deduced a new word 'disruptive technology'. After that, it became a catchword in newly created companies that aim to produce product with appeal to majority. Establishing companies with limited resources could seek for technology through contriving a completely new way of achieving an outcome. The main target of established companies is to achieve best outcome and endeavor incremental growth instead of revolutionary changes. They satisfy the demand of their customers.

### 2.2.2 The Main Elements of Disruptive Technologies

Disruptive innovation is one of the four kinds on innovation matrix that involves sustaining, disruptive, incremental, and radical. Initially, disruptive innovation is insignificant compared to existing goods and service in the market examined by traditional value metrics. The initial introduction of disruptive innovation into the market can serve only a small and commonly low-yielding customer group, but established companies are concentrated on rending more exacting, high-end customers.

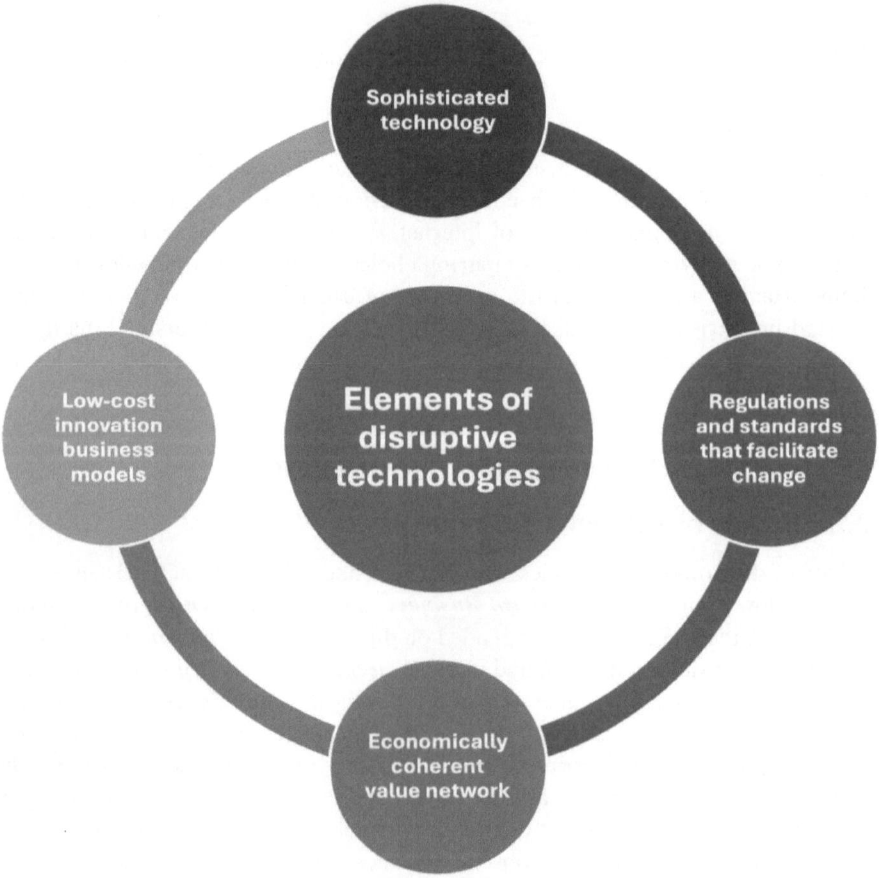


**FIGURE 2.2** Elements of disruptive technologies.

Disruptive technology is defined as a technological innovation that provokes disturbance into a business or industry. In case of adoption of disruptive technologies, incumbent businesses are frequently obliged to reconsider and re-examine their approaches or adjust their business approaches to compete with new market players. This occurrence is called technological disruption. The elements of disruptive innovation are presented in Figure 2.2. The elements are advanced technologies, a low-cost innovation business model, economically coherent value network, and regulation and policies to simplify the change. Advanced technologies and sophisticated technologies are the main pillars of disruptive technologies. They should make a product more efficient and more affordable, and accessible for wider customers. In terms of regulation and standards, they are required to facilitate the change caused by adopting disruptive technologies that can rise regulatory challenges since regulation usually responds to facts and currently, we are handling with data, thus, regulations should be anticipatory and actively responsive to data. Economically coherent value network is another element of disruptive technology, and it presents an ecosystem where all company business collaborators, distributors, and consumers avail from the disruptive technologies.

### 2.2.3 The Main Characteristics of Disruptive Technologies

The fundamental and distinguished characteristics of disruptive technologies are displayed in Figure 2.3. There are eight main characteristics of disruptive technologies and are described in detail in this paragraph.

- Targeting Underserved or Non-Consuming Markets: Disruptive innovations usually commence by aiming specialized or insufficiently served market segments that are unmarked or not properly handled by available solutions. They can serve customer with a previous inability to access or afford some goods or services.
- Lower Cost and Accessibility: The main benefit of disruptive innovation is providing reduced cost and higher accessibility in comparison to available solutions. They support technology development and efficiency enhancements to offer low-cost and more-accessible alternatives.
- Initial Performance Trade-Offs: Disruptive innovations can firstly offer lower efficiency or aspects in comparison to existing goods and service. Also, they cannot satisfy the needs of regular clients or be appropriate for all-use cases. While they usually advance quickly during the years, tightening or exceeding the efficiency gap.
- Technological Advancement and Iterative Improvement: Disruptive innovations depend on technological development or new approaches to deal with market demand. They experience progressive enhancements, and learning to boost their abilities, efficiency, and whole-value suggestions.
- Market Displacement and Industry Transformation: Disruptive innovations can interrupt available market and industries. They may defy founded player, business models, and technologies, resulting in shifting in market

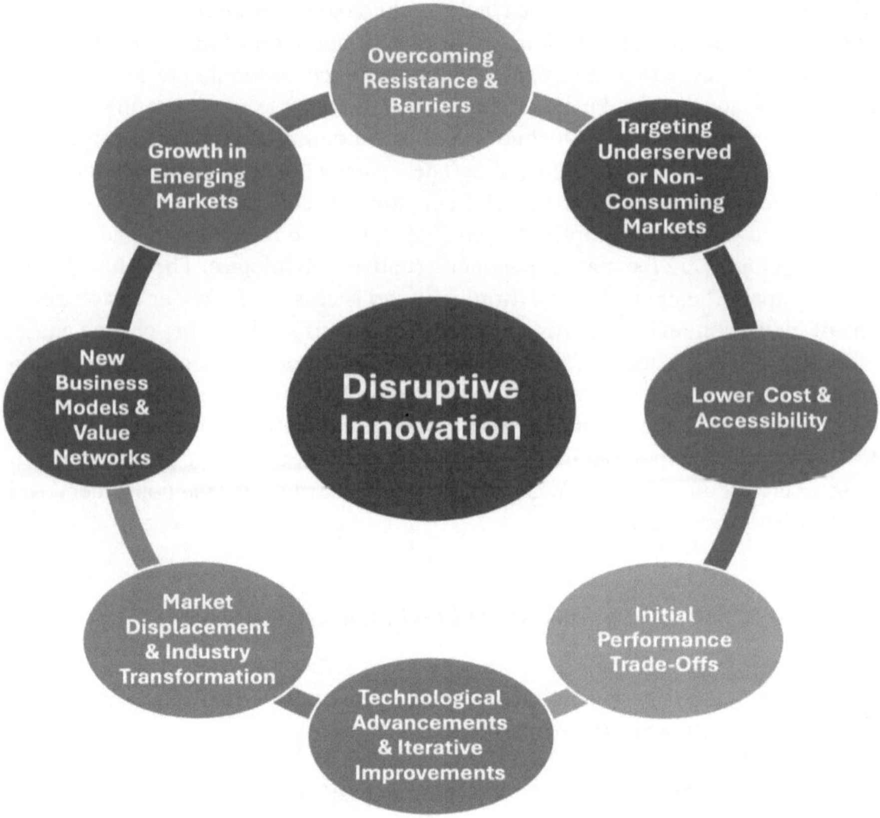


**FIGURE 2.3** The main traits of disruptive technologies (Copyright © 2024 IdeaScale) [3].

dynamics. Incumbents could face difficulties to acclimate, but newcomers exploit on growing prospects.

- New Business Models and Value Networks: Disruptive innovations usually lead to business model innovation or value creation chain that is not the same from traditional technique. They can include alternative distribution chains, node-to-node tribunes, or innovative revenue models that vary based on the way of establishing, supplying, and acquiring the value.
- Growth in Emerging Markets: Disruptive technologies usually detect initial attraction and increase in developing markets, in which the conventional infrastructure or solutions are insufficient. They may supply revolutionizing solutions in areas with restricted access to existing goods or services.
- Overcoming Resistance and Barriers: Disruptive technologies encounter opposition from incumbent players, existing practices, and available mentality. Facing resistance needs efficient communication, stakeholder coordinator, and settling issues regarding feature, accuracy, or existing norms.

### 2.2.4 Impact of Disruptive Technologies on Different Industrial Sectors

Historically, innovation in technologies achieved a powerful enhancement in productivity, remodeling industries, and putting societies on new way to expand. For instance, the growth of wireless technology essentially modified and improved telecommunications. Also, consumer-electronics industry was transformed by smartphones and PCs, enhanced by ever-smaller and quick chips. Energy markets are on the edge of an analogously radical transformation. The growth and development in many areas including semiconductors, software, and consumer electronics faced the high rise of cost of oil, aluminum, steel, and other commodities by significantly improving the way of consuming and producing energy.

Currently, digital disruption positively affects different industrial sectors:

- Academia: Currently, digital transformation has become a preference for higher education institution. Digital learning technologies are projected to improve higher education. According to the European Commission, digital education will simplify more personalized, adaptable, and student-centered teaching [4], [5].
- Communication—Telephone Mobile: Mobile communication enables a multimodal communication between people from anywhere to anywhere at any time, affecting the life at home, work, and school. Industry predicts a growth of smartphone owing to the development of 5G and forecasts that smartphone shipments will exceed 1,500 million units by 2050.
- Automobiles (Rail Transport): Disruptive technologies made a revolution in transportation sectors. Different technologies are used, including internet of things (IoT), big data, artificial intelligence (AI), and drones. The IoT allows people, process, and data to be connected through sensors and controlled remotely through the internet. AI also plays a significant role by enabling machine to operate as human brain. Moreover, the use of big and complex data with high volume has a potential in solving business problems and makes informed decisions. Additionally, drones are considered as aerial vehicles or flying robots.
- Medical—Radiography (X-Ray Image) Ultrasound: Disruptive technologies enable a fast and precise detection of abnormalities, starting from determining tumors during radiological examination to detecting early signs of eye disease. AI applications are used on medical imaging, involving cardiology, radiology, and pathology. Additionally, the use of AI in diagnostic process accelerates the interoperation of complex image and enhances early detection of disease, offering better results for patients. Moreover, AI-based image processing simplifies personalized treatment plans, leading to an optimization of healthcare delivery.
- Banks: Disruptive technologies allow us to do around 80% of our bank operation on computer or on mobile phone. This can save time and effort since we don't need to visit the bank.
- Construction: An efficient planning and optimizing use of resources positively affect the construction sector through advanced disruptions.

For instance, the use of new effective materials, 3D printing and virtual reality, has achieved huge success in increasing productivity and ensuring a stable construction process.

- Digital Photography: Disruptive technologies provide high-pixel cameras that consume lower power.
- Publishing Printers: They provide high printing quality with lower cost.
- Music and Video: Disruptive technologies ensure the production of high-quality instruments with lower cost installed in mobile phone or gadgets that can substitute the camera demand for individual use.

## 2.3 TECHNOLOGIES USED FOR ENERGY SAVING IN DIFFERENT INDUSTRIAL SECTORS

Disruptive technologies are known as innovations with an ability to launch new processes, techniques, and market that disturb existing practices and distribution system. These technologies have the potential to achieve enhancement in productivity and performance, resulting in creating new business opportunities, relieving environmental footprint, and social wealth. The main disruptive technologies that benefit the energy efficiency in different industrial sectors are presented in this part.

### 2.3.1 Energy-Management System

It is defined as a set of practices that aims for a continuous enhancement of energy use and operating costs. A set of measurable equipment is used to track energy use, monitor energy bills, and optimize energy performance. Additionally, some activities related to renewable energy and promote its use are involved in Energy-Management System (EMS), such as establishing fiscal projections for assigning renewable energy services. The implementation of EMS is based on using technologies such as utility-tracking software to forecast the future energy demands, and to scheme power budget. This can help decision makers of companies in ensuring the correlation of its strategy of energy management with its goals and fiscal scheme. The use of IoT, big data, and advanced connectivity plays a crucial role in analyzing energy data and dealing with challenges related to energy consumption and energy management. EMS can minimize the carbon footprint and our reliance on increasingly restricted fossil fuels. The existence of energy-management software in companies facilitates the management of fiscal plan and minimizes the threat linked to the rise of energy cost that negatively affected the operation of the business. An efficient budget and a more complete insight regarding the operational cost could be achieved using an energy efficiency and financial tracking for utilities. A reduction of 10% of energy use could rise the net operating income by 1.5%, as stated by Energy Star. The data collected by traditional building management system does not give a clear image on the wasted energy usage. In contrast, EMS provides a systematic, credible, and customized energy reports. The infrequent conduction of scheduled checks causes a waste of money and time, and any unexpected breakdown of equipment leads to unexpected expenses. However, intelligent EMS provides an immediate alert in case of an equipment breakdown and energy wastage. Additionally, conducting a proactive maintenance with a preventive maintenance schedule ensures the operation of the equipment and a rise of their lifetime. The use of EMS provides an in-depth energy data resulting in making smart decisions in terms of energy retrofits and adopting initiatives for cost saving.

### 2.3.2 Advanced Analytics

Advanced data analytics techniques, like AI and machine learning (ML), are frequently used to enhance energy management in smart cities. Several AI and ML solutions could be used to optimize the integration of renewable energy sources and emerging technologies like energy storage solution into the electricity system of smart cities, contributing to more sustainable and resilient cities with enhanced life quality. Advanced sensor and communication network are used in smart grids to collect and analyze data regarding electricity consumption, generation, and distribution. The collection and the analysis of data will be employed to optimize the electricity use and distribution in the city, and minimize the demand for fossil fuel power plants, which are expensive and cause pollution. Several tasks, such as load forecasting, distribution network management, fault detection, failure detection on power system equipment, and demand-side management, could be achieved by using AI and ML algorithms. Additionally, forecasting and controlling the effect of distributed energy resources on the electric grid could be conducted by AI and ML algorithms. For instance, several ML algorithms, like deep neutral networks, support vector machines, gradient boosting machines, and random forest, can be applied to forecast the yield of solar modules and wind turbines, granting network operator to enhance the balance between demand and supply. Regarding AI algorithms, ANN, which is a computational model inspired by the way biological neural networks in the human brain process information, could be utilized to recognize patterns in energy consumption and forecast the forthcoming energy demands, enabling the operator to preventively manage the electric network and avoid power cut [6].

The integration of emerging technologies related to energy storage systems and vehicle to grid (V2G) plays a fundamental role in enhancing the energy management in cities. In the V2G technology, the electric vehicle can offer several grid services in terms of frequency regulation and demand-side management, and grid optimization. Thus, a balance between supply and demand could be achieved, and the demand for pollution-causing carbon emitting power plant could be minimized. Moreover, V2G systems can perform as distributed energy resource and offer ancillary services like spinning reserve, voltage regulation, and load balancing [7].

The analysis of electricity consumption patterns and the driving attitude of electric vehicle could be performed by advanced analytics. For instance, a forecasting of future energy demand can be obtained by the determination of electricity consumption patterns. This can support the operators in managing the grid and in avoiding blackout. Also, a determination of energy use and an optimization of the charging and discharging of electric vehicles (EVs) could be accomplished by using data driving attitude of EVs [8].

Energy storage performs the function of enhancing energy management in smart cities. Since the excess of energy generated by renewable resources could be stored to be used when need, dealing with the intermittency of renewable energy resources and facilitating the incorporation of renewable energy into the grid. There are several available technologies of energy storage, such us pumped hydro storage system, compressed air energy storage, thermal energy storage, and batteries. These technologies offer many benefits to the grid in terms of reliability and stability. Additionally, an integration of energy storage with electric vehicles and distributed energy resources is possible, and leading to a smart, resilient, and affordable energy system [9].

### 2.3.3 Smart Grid

A smart grid is defined as an electricity network that can monitor and manage the transfer of electricity generated from different resources to match the changing electricity needs of consumers based on digital and advanced technologies. It can synchronize the requirements and the capacities of all generators, network operators, consumers, and electricity market stakeholders to function all components of the system efficiently, thus, reducing costs and environmental footprint, and enhancing reliability, flexibility, and resilience of the system. Smart grid involves the use of the newest sensors and computerized algorithms for all features of the function in a way that the entire system is on autopilot.

A microgrid represents a small-scale utility grid that functions independently or in association with the main grid. It is a small power-supply system that combines distributed energy resources, like solar panels and back-up generators. The main function of the microgrid is to provide continuous power supply to a particular region. An easy disconnection of microgrid from the main grid can be performed in cases of emergencies or for electrical power optimization. The microgrid has many advantages in terms of reliability, improved efficiency, and integration with renewable energy resources [10].

The value of smart grid could be enhanced with the use of evolving technologies, for instance, the combination of solar panels with wind energy, and plug-in electric vehicle.

Inserting smart grid to the electrical power grid has many benefits. It can enhance security through minimizing the effect and likelihood of environmental disaster and man-made disaster [11]. Also, it can minimize the mortality and injuries from utility-grid-associated events. Smart grid enables the transportation sector to be involved in the energy transition, since electric vehicles could be integrated in the smart grid to generate and store energy. Additionally, smart grid has the potential to boost the installation of more renewable energy resources and facilitate their penetration in the grid, minimizing the greenhouse gas emissions. Moreover, smart grid can minimize loses and energy wastes, enhancing the overall performance. Smart grid allows the energy consumers to get more choices regarding the supply and information and thus maintaining more affordability for consumers. Energy consumer can benefit from smart grid by receiving extra saving on tier energy bills. For example, consumers with central air conditioning system can benefit from placing a remote-control switch on the air conditioner by some utilities to cycle air conditioner on and off

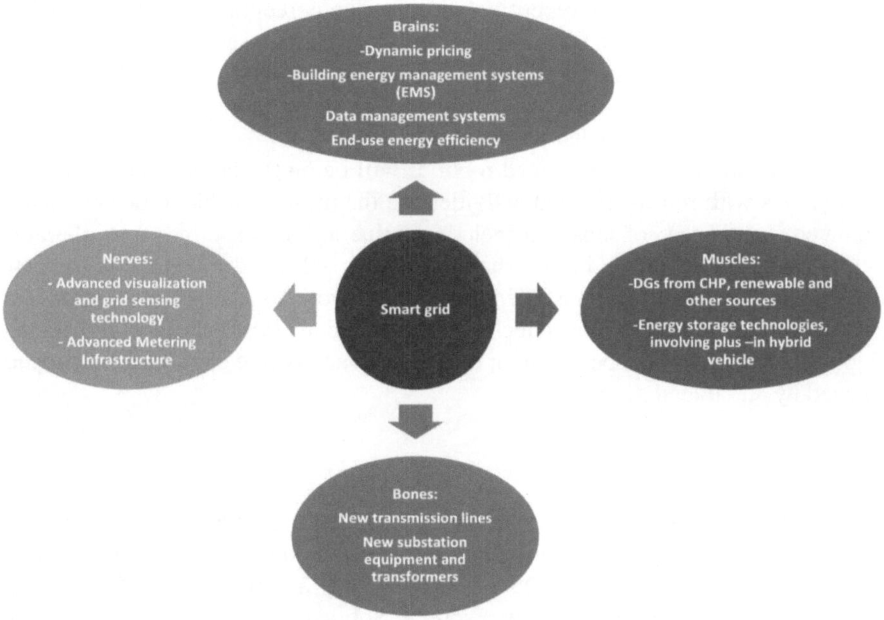


**FIGURE 2.4** Smart grid components.

during maximum energy demand. Smart grid enables such programs to function in advanced manners, leading to bigger power savings with less bothersome to proprietors and businesses. Some programs include compensation for plug-in electric vehicles, time-of-use pricing, and net metering.

The different components of smart grid are displayed in Figure 2.4 and are classified into brain, nerves, bones, and muscles.

### 2.3.4 Robotized Production System

Many benefits could be yielded by advanced robotics. A notable enhancement of productivity could be achieved by automating manual tasks since advanced robots can undertake manually performed tasks. Advanced robots could adapt to the changing process parameters, resulting in flexibility boosting by removing 'micro-stops' that commonly happen in conventional robotics process. In terms of setting up and configuration and in the case of existing a convenient simulation software, advanced robots are considerably simplified than conventional robots, and they are quick learner in performing tasks. The feature of fast ramp-up is fundamental for the production systems that need regular settlement, in retaliation to product conversions or adjustment needs. The quality of tasks could be improved by using advanced robots due to their abilities to outperform labor force in some functions, including assemble, and convey high reliability and accuracy.

Advanced robots can ensure safety by doing certain functions that are dangerous, or functions that are needed to be operated in harmful environments or functions that could cause work-related repetitive injuries. The perception of robots, which is enhanced by machine-vision technologies, could strengthen security even in unprotected environments. Advanced robots can play a fundamental role in modeling new production system that satisfies the increasing demand for more product adjustment, specialized item, and remodeled products.

The significant effect of advanced robotics will be on the labor force. More specifically, jobs with routine manual activities are the most probable to be completely automated. In the case of jobs that include routine and non-routine tasks, the non-routine tasks will achieve a higher share. Laborers will need to get more sophisticated capabilities with the transferring of manual work to non-routine works. The assigned jobs to humans will need technical skills and leadership traits. For instance, the job of technicians will be based on coping with errors that cannot be treated and managed by automated systems.

### 2.3.5 Variable Speed Drives

Variable speed drives (VSD), known also as variable frequency drives, are devices that can change the speed of a normally fixed speed motor. They are effective tools for the energy manager. The VDS system includes a three-phase AC induction motor and a variable frequency power source. Solid-state components are used by the variable frequency power source to generate a pulse-width-modulated current that changes the power and frequency delivered to the motor. A precise control of the motor speed for a wide range can be achieved because of this configuration. Additionally, the change of the speed of pump or fan according to the request with a significant saving in power consumption can be achieved by connecting VDS to fan and pump applications. VDS are more performant in terms of energy efficiency than other devices, such as inlet vanes and discharge dampers, and generate less noise at part load, enabling the functioning of fans at lower loads, avoiding the fan to function in surge, and minimizing wear on mechanical components like belts and bearings.

### 2.3.6 Light-Emitting Diode

Light-emitting diode (LED) technology, known as LED, causes a revolution in the electronics sector. This technology has continuously beheld double-digit development, and technical progresses have broadened their serviceability and abilities, resulting in enhancing its efficacity. They can perform different functions in all types of devices. They are used to form number in digital clocks, illuminate watches, transfer information from remote controls, and update you regarding the functioning of your appliances. Images could be formed on a jumbo TV screen using LEDs. The components of LEDs are tiny light bulbs that suit easily into an electrical circuit, without having filaments that burn out. The functioning principle is based on moving the electrons in a semiconductor material. There are two main categories of LEDs: visible LEDs and invisible LEDs. The main uses of visible LEDs are visual displays, switches, and enlightenment aims without using photo sensor. However, invisible LEDs are utilized in applications involving visual switches, analysis, and optical data transmission, with using photo sensors. The LEDs that are formed on gallium arsenide substrate emit red-colored light, and the LEDs that are built on gallium phosphide substrate emit green-, orange-, and yellow-colored lights. LED technology has many advantages; it has a small-size chip, a long lifetime, a low temperature, high switching speed, and high luminous intensity. Additionally, LED is eco-friendly, flexible in design with focusing light on a particular direction, high energy density, cheap, and less influenced by damages. LED is more resistant to vibrations and thermal shock, and it radiates less heat. LED bulb can last up to 25,000 hours or almost three years of constant use. On the other hand, LED technology faces some disadvantages in terms of sensitivity to damages by any excess of current and/or voltage. The radiant output power and wavelength of LED affect ambient temperatures. Also, the theoretical net efficiency can be achieved under specific cool or conditions.

The television backlighting drives mainly strong short- to mid-term market growth for LED technology, and the future mid- to long-term growth will be directed mainly by lighting applications. To enhance the LED technology, a potential for enhancement of solid-state lighting is steadily extending.

The luminous efficacy is defined as the proportion of luminous flux to the electrical input energy provided to the diode. The reaction of the eye to different wavelengths of light is known as luminous flux. The luminous efficacy determines the classification of LEDs.

LED technology is applied in different areas. It is used in homes and businesses for general illumination, decorative lighting, and functional lighting, owing to their high energy density and design flexibility. LED is used also in automotive area as headlights, interior lighting, and taillights, thanks to its long lifetime, vibration resistance, and portable size. In the areas of street and outdoor, LEDs are perfect for streetlights, outdoor signboard, and monitor, and for traffic signals. LEDs are used in displays and screens, such as computer monitors, TV, and outdoor displays for lively color, energy efficiency, and high resolution.

The newest LED version is organic LEDs, known as OLEDs, which are made using organic materials. The benefit of OLEDs is the flexibility of the used organic materials, enabling scientists to manufacture bendable lights and displays. OLEDs have been already used by manufacturers to produce smartphone and other devices.

### 2.3.7 Absorption Chiller

Chillers are employed to supply refrigeration, air conditioning, and process fluid cooling to commercial building properties and industrial installations. There are two kinds of chiller cycles: vapor compression and sorption. In the case of vapor compression chillers, centrifugal, reciprocating, and screw compressors are used to run the cycle. These compressors are usually performed by electric motor, beside steam turbine or natural gas engines. However, in the case of sorption chillers, thermal energy is generated directly by fired burner incorporated with the chiller or by thermal energy inserted indirectly to the chiller. Sorption chillers exist as adsorption or absorption design. A binary solution of a refrigerant and an absorbent is used in absorption chillers, and the cooling needs of site could be met by several options. For instance, water/lithium bromide (refrigerant/absorbent) is usually used for conditioning that needs chilling fluid temperature of 40°F or more, whereas ammonia/water is usually used for lower temperatures. Absorption chillers are the best option in terms of cost effectiveness for applications that require year-round cooling loads, such as hospitals, college campuses, hotels, and big commercial office buildings. The absorption cycle resembles the vapor compression cycle with a thermal compressor system including solution pump, absorber, and generator to replace electric motor and compressor.

A solid material is used in adsorption chiller as sorbent in the place of liquid solution. The adsorption chiller includes four heat exchangers, two adsorption chambers, an evaporator, and a condenser [12]. One of the adsorption chambers will play the role of adsorber and the other play the role of desorber in every cycle. Then, the chamber transforms their task after the saturation of sorption material with water vapor in the adsorption compartment to a certain degree, leading to the generation of a sustained cooling effect in the evaporator. The key advantage of adsorption chillers drives from its ability to be operated at low-temperature heat of 50–90°C, which can be waste heat or heat obtained from geothermal or solar energy [13]. The needed heat is used to operate pumps and automation systems. Adsorption chillers could generate synchronously chilled water and desalinated water. The chilled water is generated by removing heat from adsorption chillers in the evaporator, and the desalinated water is produced through evaporating the water twice, one in the evaporation, followed in the adsorbent bed, and then condensing the pure distilled water in the condenser. The cooling by adsorption method is considered as a sustainable method without any carbon footprint impact since the chilled water is generated without using fossil fuels. A thermal compression is used in the cooling cycle of adsorption method, resulting in several benefits that involve the existing reduced number of driving pieces, and thus less vibration, high reliability, and less noise emissions [14, 15]. Additionally, adsorption cooling systems use natural refrigerant as a substitute of chlorofluorocarbons or hydrofluorocarbons. On the other hand, this technology encounters some difficulties deriving from its operation, involving a low specific cooling power of up until 200 W/kg, and a low coefficient of performance in majority of cases up until 0.6 [16].

### 2.3.8 High-Efficiency Compressor

Air compressors play a crucial role in industry sector. They are employed in manufacturing to run pneumatic tools, drive, and control devices, and conduct quality control tests. Also, air compressors are used in food-processing industry to perform some tasks related to packaging, bottling, and handling products. Additionally, air compressors are used to lead surgical tools and to provide compressed air to medical appliances.

There are different kinds of air compressors that are suitable for different applications. For instance, reciprocating air compressors are considered as an ideal choice for irregular usage in small-scale operations, and rotary screw compressors are ideal for a continuous use in large-sale operations. For high-speed applications, such as chillers and turbochargers, centrifugal compressor is an ideal candidate. The main innovation in improving the technology of air compressor is related to energy efficiency. Several models to reduce the energy consumption and the operational costs have been developed by manufacturers. For instance, VDS are used to adapt the motor speed in conformity with air need, resulting in a considerable reduction of energy consumption during times of diminished demands. Moreover, waste heat produced during the compression process is captured by heat recovery systems to be reused for water and space heating for additional improvement of the total energy efficiency.

Smart controls and automation play a crucial role in the operation of air compressors. Intelligent control system could optimize effectiveness and energy efficiency of air compressors, supervise, and regulate the function of compressors in real time based on different factors. For example, a detection of implied issues could be performed using a predictive maintenance algorithm, resulting in reducing downtime and fixing expenses, and leading to breakdowns. Thus, incorporating air compressor with IoT technology led to remote monitoring and control, enhancing the benefits for customers.

Several industrial environments face issues related to noise pollution. Thus, air compressor with noise reduction was developed by manufacturers to ensure a noiseless operation. These models involve anti-vibration mounts, sound-absorbing enclosing, and passive fans. Innovative design elements are also incorporated in some models to minimize noise at their origin. Scroll compressors are one of the options that are progressively selected for applications that require lowest level of noise, like medical and dental offices.

### 2.3.9 High-Efficiency Motors

Motors are the principal components of the drive system of consumer and industrial goods. The application of motor system transfers/fetches the appliance of automation to mankind. Actually, motor-based systems are responsible on the consumption of more than 60% of the total energy consumption [17]. This shows the importance of improving the efficiency of the motor system energy in order to achieve the total national final energy conservation through developing high-efficiency motors. Also, the development of motor efficiency specifications by global organization is a rising trend with direct implications on the generation and design of motor. The replacement of traditional motor by more efficient motors will offer many benefits to the environment and effective use of resources, along with fabrication cost and in competitiveness. The traditional motors will be removed with the boom of the era of high-efficiency motors. An enhancement in the energy efficiency of motors has achieved high levels, categorized as super and ultra-premium efficiency motors (IE4 and IE5). Actually, induction motors with IEA efficiency are completely available to customers [18]. Permanent-magnet motors and synchronous reluctance motors enable motors to surpass the IE4 and IE5 efficiency limits [19]. Three-phase induction motors are yet the preferred option owing to their rigidity and accuracy for fixed-speed applications. Regarding the technologies that are competent with starting direct on-line (DOL), Line-Start Permanent Magnet Motors and DOL Synchronous Reluctance (SynRel) motors are also available on the market, but they face some operational challenges that affect their range uses [20]. On the other hand, motors used for speed variation application, such as Permanent Magnet Synchronous Motors and SynRel motors, could be an effective and workable alternative to induction motor, competing in reliability and dominating in performance. These kinds of motors perform at synchronous speed, eradicating losses in the rotors, resulting in attaining significant performance levels that can surpass IE5. Additionally, a great reduction in the total size and weight of the motor and thus less use of materials for its fabrication could be achieved by enhancing the properties of these technologies [21].

### 2.3.10 High-Efficiency Pumps

Buildings account for a significant percentage of the global energy utilization. A maximization of heating and cooling efficiency leads to minimization in the total energy utilization by 30–40% [22]. Heat pump is one of the effective technologies that can be used for an efficient heating and cooling. It is used to transfer heat between a cold and a hot space. In summer, heat pump transfers heat from inside to outside, while in winter, it drags heat from outside and transfers it inside. The heat sink could be air, water, and the ground. According to IEA [23], buildings account for 4 gigatons of $CO_2$ emission yearly, around 10% of global emissions. Heat pump could be eco-friendly if the electricity used to power it is generated from renewable energy resources, enabling houses and manufacturing facilities to cut their carbon emissions. High-efficiency heat pump could be sustainable and efficient for interior space conditioning. Currently, the existing heat pumps are three to five times more performant than natural gas boilers [23]. In order to minimize the building energy cost without major investing or making structural adjustment, retrofitting existing HVAC systems, that used to regulate indoor environments by controlling temperature, humidity, and air quality, is a feasible option. Some changes or some addition to the existing system could be needed, but a considerable energy saving could be achieved.

Heat pumps can also meet the heating demands needed in industry and district heating. Currently, large heat pumps can offer heat up to 140–160°C, with larger temperature possible through creativity and enhanced design. The functioning principle is derived on the thermodynamic of the vapor compression refrigeration cycle, in which heat pump functions in a closed cycle with a refrigerant fluid revolving within. The refrigerant fluid experiences a conversion from liquid to vapor state under constant pressure in the evaporation, absorbing heat in the process. After that, the temperature and pressure of the refrigerant is increased by using a compressor, to prepare it for the condenser.

### 2.3.11 Immersion Cooling Technology

Immersion cooling method is used to cool data center IT hardware batteries, solar photovoltaics (PVs), and power transformers, through a direct immersion of hardware in a non-conductive liquid. This method is based on using a dielectric liquid to soak up heat source by convection and liberate to the environment via a single-phase or two-phase system [24]. The advantages of immersion cooling derive from its high heat transfer coefficient, steady hydrodynamic flow, and the capability to use liquid for a direct cooling of hot components [25]. Mineral oil and virgin coconut oil are the two main dielectric used. The ability of immersion cooling to cool the whole area leads to the enhancement of temperature uniformity through minimizing the impact of local heating on the negative and positive electrodes [26]. The use of cooling medium that offers high thermal cooling, high heat capacity, and low viscosity could enhance the efficiency [27]. On the other hand, there are some issues related to electric short circuit and electrochemical corrosion. The main criteria that should be considered in the selection of the dielectric material are good chemical stability, non-toxicity, non-flammability, preventing an electric short circuit, and safety concerns [28]. Regarding the cooling medium, silicon-based oil, deionized water, and mineral oil are the best options. Since valves, fans, and connectors are not needed in this system, energy efficiency is achieved using immersion cooling. Immersion cooling offers more benefits in terms of space required to install it compared to traditional cooling techniques, resulting in minimizing the ownership cost. Also, the installation and maintenance of immersion cooling technology are easy owing to the use of non-flammable and non-hazardous liquids, and the simpler construction design. Moreover, immersion cooling technology is not noisy and does not release greenhouse gas emission; thus, any surplus of energy to activate the muffler and emission capture is not required. The stable and uniform temperature offered by immersion cooling technology maintains the same level of performance.

### 2.3.12 Liquid Desiccant Systems

Liquid desiccant systems attract attention owing to its low regeneration temperature as well as its capability of soaking up organic and inorganic pollutants from the air. There are many applications that use the liquid desiccant dehumidifiers in air conditioning systems and in industries. Additionally, they are used to improve the quality of products for crop solar drying, and the results indicated a reduction of the cracker seeds by 7% under low temperatures (between 45°C and 55°C) [29]. The widely used liquid desiccant solutions are $CaCl_2$, LiCl, and LiBr, and the results indicated a high performance of LiCl liquid desiccant compared to the other solutions [29]. The main benefit of liquid desiccant technology is the non-condensation on interior surface of the equipment, resulting in removal of the growth of several fungi and bacteria and this prevents several health problems. It is beneficial to different sectors, mainly food storage and medicines. Also, the regeneration of the liquid desiccants could be performed under low temperature, contributing to the feasibility of low-grade energy sources. Moreover, concentrated liquid desiccant could be stocked and used in case of unavailability of heat source at some time. The whole unit is compact and small; it minimizes the use of electrical energy with high capability to remove moisture. Liquid desiccant materials are good antifreeze, preventing freezing or frosting at lower temperature [29]. There are some drawbacks of liquid desiccant materials despite their numerous advantages. The liquid desiccant is corrosive and suffers from crystallization issue that can negatively affect the system. The excess of liquid desiccant with air treatment could affect occupant's health. The cost of the system is expensive, and it requires a large pump to pump large volume of liquid desiccant solution.

**TABLE 2.1**
**Estimated impact and maturity level of disruptive technologies for energy efficiencies in industrial sector**

| Technology | Estimated impact | Maturity level |
|---|---|---|
| Energy management | 2–5% | Commercial |
| Advanced analytics | 2–5% | Commercial |
| Smart grids | 2–5% | Commercial |
| Variable speed drives | 2–5% | Commercial |
| Light-emitting diode | 2–5% | Commercial |
| Absorption chiller | 2–5% | Commercial |
| High-efficiency compressor | 2–5% | Commercial |
| High-efficiency pumps | 2–5% | Commercial |
| High-efficiency motor | 2–5% | Commercial |
| Robotized production system | 5–15% | Commercial |

## 2.4 ESTIMATED IMPACT AND EFFECT ON ENERGY SECTOR OF DISRUPTIVE TECHNOLOGIES FOR ENERGY EFFICIENCIES

The estimated impact and the maturity level of the disruptive technologies are presented in Table 2.1. All the presented disruptive technologies achieved a commercial level with an estimated impact of 2–5%, except the robotized production system, that achieved an estimated impact of 5–15%.

The capabilities and the effect on energy sector for the disruptive technologies presented in this chapter are displayed in Table 2.2. These technologies can enhance the overall efficiency of the system, minimize the energy consumption, and promote the integration of renewable energy. Some technologies like absorption chiller and liquid desiccant systems can be powered by waste heat or renewable energy.

## 2.5 CONCLUSIONS

Disruptive technology is defined as a technology that transforms the operation method of business. In the case of effective implementation, it will replace the existing system owing to its unique attributes. The incorporation of energy, with materials, labor, and information, into productivity processes leads to improved technology.

Energy accounts for a significant part of company-operating costs. On a global scale, energy accounts for around one-third of operating expenses in the cement, mining, chemical, and metal sectors. Several disruptive technologies offer significant benefit for energy saving in different industrial sectors, including EMS, advanced analytics, smart grid, LED, robotized, high-efficiency pump and motor, high-efficiency compressor, liquid desiccant system, immersion cooling technology, absorption chiller, robotized production system, and VSD. This chapter presents an overview on these disruptive technologies. All the presented technologies offer a reduction

**TABLE 2.2**
**Capabilities and the effect on energy sector for the disruptive technologies used for energy efficiency**

| Technology | Capability | Effect on energy sector |
|---|---|---|
| Advanced robotics | Productivity enhancement and automation | Rise of electricity demand, and rise of energy needed for management due to the additional equipment |
| Smart grid | Enhancement of efficiencies, rise flexibility and reliability of the grid | Uplifter capacity for using renewable energy, maximize the management of energy production |
| Energy management | Enhance the productivity and improve product quality | Reduce energy usage, promote clean energy sources |
| Variable speed drives | Enhance the energy efficiency of the system by optimizing the power supplied to the system, minimize motor losses | Energy saving, enhance system efficiency, better energy management |
| Absorption chiller | Can be driven by renewable energy resources, can use waste heat from industrial process, use thermal energy and this minimizes energy demand, enhance overall energy management | Energy saving, improving the energy efficiency of HVAC systems, low energy consumption, support renewable energy integration since it can be powered by renewable energy (solar thermal collectors) |
| High-efficiency compressor | Minimize energy wastage during the compression process, improving total performance and reliability owing to used materials and design, and enhance productivity of industrial processes | Lower energy consumption, support renewable energy generation (minimize the total energy demand) |
| High-efficiency motor | Ability to convert more electrical energy into mechanical energy, which could minimize costs, and achieve greater energy efficiency, and enhanced productivity | Energy saving and low energy consumption, support renewable energy integration through minimizing total energy demand |
| High-efficiency pumps | Promise energy efficiency and flexibility compared to traditional pumps, designed to operate using less energy, minimize heat generation, optimize the performance of the fluid handling system | Less power consumption and energy saving |
| Immersion cooling technology | Improve heat transfer efficiency, minimize cooling energy consumption, ability to reuse the heat | Less energy consumption for cooling and high efficiency |
| Liquid desiccant systems | Enhance dehumidification efficiency, minimize energy consumption by using waste heat or renewable energy, minimize need for mechanical cooling, energy recovery potential | Enhance energy efficiency, using renewable energy or heat waste, minimize cooling loads, enhance total performance of HVAC systems |

of energy consumption, and an enhancement of performance and energy efficiency. Some of them enhance the integration of renewable energy resources. Most of the presented technologies exist, while there are some constraints that companies face in terms of the way to choose the suitable technology, the operation way of the selected technology, and the way to renew them to maintain a continuous functioning.

## REFERENCES


1. IEA. *World Energy Outlook* 2023. 2023, Paris.
2. Harsh Chouldhry, M. L., Ken Somers, & Joris Van Niel. *Greening the Future: New Technologies that Could Transform How Industry Uses Energy.* 2015, McKinsey&Company, Singapore.
3. Jain, N. *What Is Disruptive Innovation? Definition, Examples and Best Practices.* 2023, IDEASCALE.
4. Rodero, L.G. The European digital education plan for the development of lifelong learning. *The Value of the Difference and Lifelong Learning in the Contemporary Pedagogy*, 2023: p. 113.
5. Prakash, C., R. Yadav, A. Dangi, & A. Singh. Factor analysis approach to study mobile applications' characteristics and consumers' attitudes. In *Computational Intelligence for Engineering and Management Applications: Select Proceedings of CIEMA 2022* (pp. 783–798). Singapore: Springer Nature Singapore, 2023.
6. Moreno Escobar, J.J., O. Morales Matamoros, R. Tejeida Padilla, I. Lina Reyes, & H. Quintana Espinosa. A comprehensive review on smart grids: Challenges and opportunities. *Sensors*, 2021, 21(21): p. 6978.
7. Longo, M., W. Yaici, & F. Foiadelli. Future developments in vehicle-to-grid technologies. In John R. Vacca (ed.), *Solving Urban Infrastructure Problems Using Smart City Technologies.* 2021, Elsevier, pp. 613–629.
8. Qian, T., C. Shao, X. Wang, & M. Shahidehpour. Deep reinforcement learning for EV charging navigation by coordinating smart grid and intelligent transportation system. *IEEE Transactions on Smart Grid*, 2019, 11(2): pp. 1714–1723.
9. Sami, M.S., M. Abrar, R. Akram, M.M. Hussain, M.H. Nazir, M.S. Khan, & S. Raza. Energy management of microgrids for smart cities: A review. *Energies*, 2021, 14(18): p. 5976.
10. Hartono, B., Y. Budiyanto, & R. Setiabudy. Review of microgrid technology. In *2013 International Conference on QiR*. 2013. IEEE.
11. Bush, S.F. *Smart Grid: Communication-Enabled Intelligence for the Electric Power Grid.* 2014, John Wiley & Sons, United States.
12. Sah, R.P., B. Choudhury, R.K. Das, & A. Sur. An overview of modelling techniques employed for performance simulation of low–grade heat operated adsorption cooling systems. *Renewable and Sustainable Energy Reviews*, 2017, 74: pp. 364–376.
13. Zajaczkowski, B. Optimizing performance of a three-bed adsorption chiller using new cycle time allocation and mass recovery. *Applied Thermal Engineering*, 2016, 100: pp. 744–752.
14. Fernandes, M., G. Brites, J. Costa, A. Gaspar, & V. Costa. Review and future trends of solar adsorption refrigeration systems. *Renewable and Sustainable Energy Reviews*, 2014, 39: pp. 102–123.
15. Pan, Q., J. Peng, & R. Wang. Application analysis of adsorption refrigeration system for solar and data center waste heat utilization. *Energy Conversion and Management,* 2021. 228: p. 113564.
16. Rogala, Z. Adsorption chiller using flat-tube adsorbers–Performance assessment and optimization. *Applied Thermal Engineering*, 2017, 121: pp. 431–442.

17. Hajian, M., G.A. Markadeh, J. Soltani, & S. Hoseinnia. Energy optimized sliding-mode control of sensorless induction motor drives. *Energy Conversion and Management*, 2009, 50(9): pp. 2296–2306.
18. De Almeida, A.T., F.J. Ferreira, & G. Baoming. Beyond induction motors—Technology trends to move up efficiency. In *49th IEEE/IAS Industrial & Commercial Power Systems Technical Conference*, 2013. IEEE.
19. De Almeida, A.T., F.J. Ferreira, & J.A. Fong. Standards for efficiency of electric motors. *IEEE Industry Applications Magazine*, 2010, 17(1): pp. 12–19.
20. McElveen, R., M. Melfi, & R. Daugherty. Line start permanent magnet motors-Starting, standards and application guidelines. In *2014 IEEE Petroleum and Chemical Industry Technical Conference (PCIC)*, 2014. IEEE.
21. Melfi, M.J., S. Evon, & R. McElveen. Induction versus permanent magnet motors. *IEEE Industry Applications Magazine*, 2009, 15(6): pp. 28–35.
22. Li, Y., Y. Rezgui, & H. Zhu. District heating and cooling optimization and enhancement–Towards integration of renewables, storage and smart grid. *Renewable and Sustainable Energy Reviews*, 2017, 72: pp. 281–294.
23. IEA. *The Future of Heat Pumps*. 2022, IEA, Paris.
24. Nadjahi, C., H. Louahlia, & S. Lemasson. A review of thermal management and innovative cooling strategies for data center. *Sustainable Computing: Informatics and Systems*, 2018, 19: pp. 14–28.
25. Birbarah, P., T. Gebrael, T. Foulkes, A. Stillwell, A. Moore, R. Pilawa-Podgurski, & N. Miljkovic. Water immersion cooling of high power density electronics. *International Journal of Heat and Mass Transfer*, 2020, 147: p. 118918.
26. Bashirov, M., A. Khismatullin, & E. Sirotina. Cooling system oil-immersed transformers with the use of a circulating sulfur hexafluoride. In *Advances in Automation: Proceedings of the International Russian Automation Conference, RusAutoCon 2019*, September 8–14, 2019, Sochi, Russia, 2020, Springer.
27. Xia, G., L. Cao, & G. Bi. A review on battery thermal management in electric vehicle application. *Journal of Power Sources*, 2017, 367: pp. 90–105.
28. Chemisana, D., E. Fernandez, A. Riverola, & A. Moreno. Fluid-based spectrally selective filters for direct immersed PVT solar systems in building applications. *Renewable Energy*, 2018, 123: pp. 263–272.
29. Gurubalan, A., M. Maiya, & P.J. Geoghegan. A comprehensive review of liquid desiccant air conditioning system. *Applied Energy*, 2019, 254: p. 113673.

# 3 Artificial Intelligence as Disruptive Technology in Human Resource Management

*Gulab Kumar, Dipanker Sharma, and Bhawana Bhardwaj*

## 3.1 INTRODUCTION

The business world nowadays is fast moving in all aspects. Artificial Intelligence (AI) could be at the root of innovation and competitive advantage, but at the other end, AI has caused disruptions to the present-day status norm in terms of systems, processes, and events [1]. The definition of disruption as per the Cambridge Dictionary[1] is "an interruption in the usual way that a system, process, or event works" and as a measure to stop the event, process, or system from "continuing as usual or as expected". This chapter is focused on the idea that AI is definitely changing Human Resource Management (HRM) practices in organisations. We will start by explaining what AI is and why it is important in current business contexts, and we will then give an overview of HRM and its changing role in the company's strategic objectives. Lastly, we will introduce the central theme of the chapter: AI disruption in HRM functions.

### 3.1.1 Artificial Intelligence: Its Significance in the Modern Business Environment

AI is a concept that reflects the imitation of human cognition processes by machines. It is a term that is broad and encompasses several technologies, including machine learning (ML), computer vision, and natural language processing (NLP). Varma (2023) [2] defines AI as "computer systems capable of imitating human intelligence." AI systems are capable of analysing large amounts of data, identifying trends, and making independent decisions on their own, which is similar to the human brain but done in a more accurate and effective manner.

AI today is not only an innovation-enhancing tool but also helps in increasing efficiency and making better strategic decisions. AI identifies problems and develops solutions that benefit people, society, and technology [3]. Organisations in any industry use AI to automate their routine tasks, eliminate processes, and extract valuable


DOI: 10.1201/9781032700953-4

information from data. The use of completely different AI solutions, like customer service chatbots and predictive analytics models, is completely changing the way businesses operate and the commercial environment.

### 3.1.2 Resource Management: Overview and Its Transforming Role in Organisational Achievement

Human Resource Management (HRM) refers to a wide range of activities aimed at managing an organisation's most valuable asset, its people. In the past, HRM was defined as an administrative function, but later it evolved into a strategic function that plays a significant role in achieving organisational goals. HRM operations include identification, training and development, employee relations, performance management, and staff retention.

In today's globalised, technologically sophisticated business environment, which is changing along with the demography of the workforce, HRM is facing new challenges. It is also providing plenty of new opportunities. Every product or service used by people has changed as a result of technological advancements. Businesses that are able to adapt to this transformation will survive, while others will ultimately lose market share [4]. Companies understand that to stay ahead of the attracting competition, developing and retaining the best talent personnel possible is a real priority. According to a report by George & Thomas (2019) [5], approximately 38% of all the organisations worldwide used AI for HRM tasks in 2018. Consequently, HRM is currently seen as an integral part of strategic planning in the organisation for achieving objectives as well as creating and maintaining a culture of creativity and excellence.

### 3.1.3 The Disruptive Impact of AI on HRM Practices

In 2021, Minbaeva [1] stated in her paper that AI, digitalisation, ML, and a flexible workforce are main causes of workplace disruptions, and digitisation of business models is one of the largest disruptions for business practices. By this time, AI technologies are significantly developing, and they are disrupting the way that how companies manage their employees. AI comes with new technologies for recruitment, engagement, performance management, development, and succession. The way businesses are run has been altered by disruptive forces, and human resources (HRs) need to change with the times. HR needs to design a digital workplace where people can work together, develop, and solve business problems [6]. Through integrating AI into HRM processes, businesses can automate their operations, enhance decision-making, and open up new possibilities for creativity and development. While AI integration brings these advantages, it also has a few challenges, like ethical dilemmas, data privacy considerations, and the threat of workforce displacement. This chapter is aimed at identifying the possible opportunities and challenges that are associated with AI and discussing them in detail. It will provide insights into how organisations can benefit from AI to achieve excellence in HRM and accomplish their strategic goals.

Following this (1) Introduction, the chapter is organised as follows: (2) Understanding Disruptive Technology, (3) Transforming Recruitment Processes with AI, (4) Revolutionising Employee Engagement Strategies, (5) Enhancement of Performance Evaluation and Feedback Mechanisms, (6) Leveraging AI for Talent Development and Succession Planning, (7) Ethical and Responsible Implementation of AI in HRM, and (8) Future Directions and Conclusion.

## 3.2 UNDERSTANDING DISRUPTIVE TECHNOLOGY

### 3.2.1 Emergence of Disruption Technology: Its Implications for Industries and Organisations

Disruptive technology describes innovations that fundamentally change business operations by making one of the key factors (products, services, or processes) within the business less useful or completely useless. Also, these types of technologies create brand new opportunities for company's growth and effectiveness. These technologies adopt the concept of creative destruction and disruption in existing categories of industries, where previously companies were kings, and now the market is changing its shape. On the one hand, it proposes unexpected opportunities for innovation and differentiation, and at the same time, they can be seen as a challenge to competitive advantage on the other. The disruptive aspect sometimes appears to be dangerous as it poses a threat to incumbents who are left behind, which can result in consequences including loss of market share in the form of market decline and organisational extinction. HR professionals are warned by several consulting studies to be prepared for disruptions, including the digital revolution, automation, and AI [1].

The impact of emerging technologies is far more far-reaching than just individual enterprises; across industry lines, they completely change the supply and demand chain structure, business models in general, and customer expectations. The rise of automation can lead to social and economic changes, resulting in changes to employment patterns, business imperatives, and accepted norms. In other words, disruptive technology is a mechanism for change that puts pressure on organisations to innovate, while those who are reluctant to change will eventually become obsolete.

### 3.2.2 Historical Examples of Disruptive Technologies and Their Transformative Effects

There are many historical examples of disruptive technologies that changed the entire traditional approach. Let us discuss some examples that show the threat of any technological innovation to existing industries and businesses (Figure 3.1).

The most impactful disruption was the introduction of the "Internet" in the field, and it became the focal point of communication, commerce, and information flow revolutions. In the beginning, the Internet was considered only a temporary fashion or fad. But it quickly changed to disrupt every business and industry model and provide diverse opportunities such as e-business, web advertising, and the distribution of digital content. Businesses that did not understand what role the

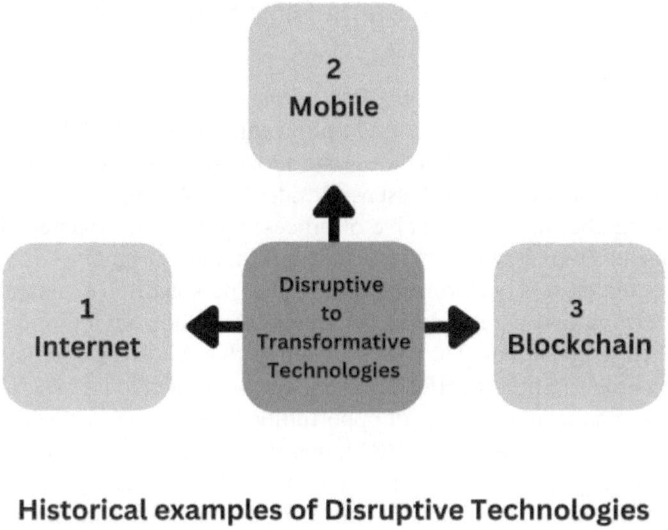


**FIGURE 3.1** Historical disruptive technologies

**Source: Created by author**

Internet could play for them were pushed to the extreme edges of the market, while those that did learn about the relevance of the Internet evolved within the digital economy.

Furthermore, one can recall that the technology of "mobile" is changing the relationship between people, work, and media based on the smart phone and its applications. The smartphone became a single device, the sector, which replaced the phone calls, media, and computing software that landlines had left behind. Companies that have adopted the Mobile-First strategy have offered new customer interaction schemes as well as an increase in revenues, thus adopting new business models. Companies that did not adopt new business models were always at risk of irrelevance, instability, and declining market share.

"Blockchain," another new technology, has not only reshaped the way many important business operations are conducted but has also entered the financial sector. Blockchain is a new financial technology that is distributed, honest, and reliable, making it disruptive to the current centralised, insecure banking and payment systems. Blockchain has introduced novelties such as value expressed in digital currencies and smart contracts that impact the profitability models of traditional companies and require the emergence of new legal frameworks. There are clear indications that organisations that actively adopt blockchain technologies can become candidates for greater efficiency, transparency, and trust than those who ignore blockchain and may lose out in the fast-moving world.

In addition to these, the profound disruptive and enabling effects of robotics, machine intelligence, and deep-learning skills on machines have impacted government, business, and society [3].

### 3.2.3 AI as a Disruptive Factor in the HRM Function of the Organisations

AI is a technology that has disrupted HRM because of the features that set it apart. The first benefit is the "automation" and optimisation of routine HR tasks, such as analysing applicants, building schedules, and processing payslips, freeing employees to concentrate on other critical issues. Besides that, AI improves the "quality of decision-making" because it is capable of processing massive volumes of data and searching for patterns in it, which may include information on employee engagement, labour market trends, and performance. AI creates "personalised experiences" (such as custom learning and career plans) for employees, who can stay motivated and engaged in their job positions. The next key point is that AI offers "predictive analytics and forecasting" abilities to HR managers, allowing them to foresee future talent needs and risks and take advantage of opportunities. Hence, these new technologies are starting to be incorporated into HRM, and we are seeing a digital boom that is drastically changing our work environment and how we view HR [6,7].

In terms of HRM, AI has a high disruptive power that reflects in the improvement of human performance, the enhancement of process optimisation, and the development of innovations. AI's remarkable development has created a revolution that will likely continue for a long time [8]. On the other hand, careful analysis of the ethical, legal, and social ramifications of its implementation should be considered before the implementation, such as data privacy, mitigation of biases, and workforce displacement. Through the utilisation of AI while simultaneously managing the challenges entailed in this process, companies can realise novel possibilities for growth, competitiveness, and the overall fulfilment of their organisational objectives in the digital era.

## 3.3 TRANSFORMING RECRUITMENT PROCESSES WITH AI

Recruitment is one of the most fundamental aspects of HRM because it opens the door for companies to come in and attract the best people. AI has transformed to a large extent the recruitment processes, which were not possible without speed and precision before. AI helps HR planning by anticipating future workforce requirements and making intelligent hiring decisions [9]. Brain circulation has emerged as one of the many outcomes to this [10]. Here is a zone that talks about the way AI changes recruitment compulsorily, and a start is made by looking into traditional methods and their disadvantages. Traditional methods were unable to efficiently discover and target passive job seekers; now, AI-enabled recruiting tools are able to do so [11]. Then, AI in recruitment will be discussed, along with the advantages that it brings. Moreover, we will explore the case studies and cases where the system was used to recruit suitably qualified applicants.

### 3.3.1 Traditional Recruitment Methods and Their Limitations

The traditional methods of recruitment, for instance, job postings, resume selection, and interviews, have always been the backbone of talent acquisition systems.

Nevertheless, these methods are to some extent useful, but they are, by nature, time-consuming, labour-intensive, and, in some cases, prone to errors. Biased human screening too often either fails to find great candidates or introduces inconsistencies into the process of screening. Further still, conventional approaches would perhaps be unable to cope with high numbers of candidates in an expedient fashion, which might cause delayed hiring decisions as well as missed opportunities. Some of the recent and relevant studies recognised the necessity for established companies to embrace new technology, such as AI, in order to succeed, since they have been reluctant to do so [12].

### 3.3.2 AI Recruitment Tool Exploration and Their Benefits

AI-enabled hiring and selection are essential to attracting and hiring the best candidates for an organisation's workforce [13]. AI-powered recruitment instruments bring about a pivotal change in the way companies select and pick candidates, as opposed to the traditional ways, which have imperfections. Utilising those tools, which enable the use of NLP, ML, and algorithms to speed up and optimise the recruiting processes, lies at the core. AI in hiring not only reduces the amount of time needed to finish the procedures, but it also calculates wide range of dynamics to help with decision-making [2–14]. When it comes to AI-based recruitment systems, their essential elements comprise automated resume screening, predictive analytics, and chatbots (Figure 3.2).

**Automated Resume Screening:** AI-based resume-screening tools can understand resumes and job descriptions while identifying specific qualifications, skills, and experience that are relevant to the job. Historically, such components have the opportunity to analyse hiring data and feedback. Therefore, they can effectively match candidates to job requirements. This, in turn, reduces the time to hire and contributes to the quality of hires.

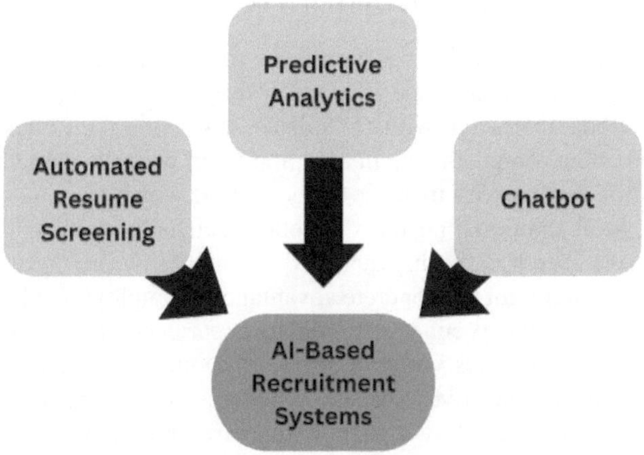


**FIGURE 3.2** Artificial Intelligence (AI)–based recruitment systems

**Source: Created by author**

**Predictive Analytics:** The AI algorithms use past records and market trends to provide forecasts about the future of hiring, candidate efficiency, and attrition rates. Through analysing various types of factors like candidate profiles, job descriptions, and external job market situations, predictive analytics equip organisations with the capability to make the appropriate data-driven decisions and recruitment strategies.

**Chatbots:** AI chatbots engage with candidates in real time and provide personalised support throughout the recruitment process. Chatbots are capable of giving information and solutions to questions that are recurrent, setting interviews and providing information about the status of a recruitment, optimising the candidate experience, and allowing the recruiters to do more demanding tasks.

### 3.3.3 Case Studies: Demonstrating Successful Implementation of AI in Recruitment

**Case Study 1: Unilever**[2]

In 2016, the consumer goods giant Unilever improved its AI-supported recruitment, selection, and onboarding processes by utilising the AI recruitment tools "HireVue" and "Pymetrics" [15]. Through the use of analytics and ML algorithms, Unilever was able to optimise its hiring procedures, improve candidate matching, and cut down on hiring time by 75%. Using NLP, these systems compare the abilities, experiences, and qualifications listed in resumes with the job requirements. The company saw a decrease in expenses and a rise in candidate satisfaction, which led to better hiring practices and higher employee-retention rates.

**Case Study 2: Hilton Worldwide**[3]

Hilton Worldwide, a hospitality corporation, integrated AI-driven chatbots to improve the recruitment experience for job seekers [16]. Through messaging platforms, these chatbots interacted with candidates, answering their questions, providing information about job positions, and helping them with the application process. As a result, Hilton saw an 85% increase in the pace at which candidates were hired. They have also seen additional business benefits, such as broadening the talent pool and facilitating the search for highly qualified candidates by their recruiters.

These instances highlight the concrete advantages of employing AI-driven recruitment tools to boost efficiency, enhance candidate experiences, and achieve outcomes. By embracing AI technologies within their talent-acquisition processes, companies can gain an advantage that drives success in the era. Recruiting technologies with AI capabilities should be used by managers who aren't already. Managers utilising these tools ought to keep learning and growing in their roles. In addition to recruiting, they want to investigate the use of AI-enabled technologies for applicant screening, evaluation, and interviews [16].

## 3.4 REVOLUTIONISING EMPLOYEE ENGAGEMENT STRATEGIES

HRs given to employee engagement are a major driver of organisational success, as they affect productivity, employee retention, and overall organisational performance. AI-enabled chatbots allow workers to speak up, which in turn affects worker engagement and enhances the idea of AI-mediated social interchange in the HR environment [17–18]. This part will analyse the employee engagement perspective within the context of AI-driven sentiment analysis tools that are redefining the way employee engagement is measured and the use of personalised learning platforms to facilitate employee engagement and retention.

### 3.4.1 Importance of Employee Engagement for Organisational Success

The term "employee engagement" indicates the level of emotional commitment the employees have towards their workplace, the organisation's objectives, and the purpose of their work. Those employees who embrace the organisation are more likely to be motivated and productive, and they always care about organisational success. They don't just do their job; they take on extra responsibility and have excellent ideas that improve the culture of work. As a result, while disengaged workers might indicate a lack of morale, absenteeism, and staff turnover, it may also lead to a lack of productivity and rising costs for organisations.

Research conducted by researchers always reveals the association between employee involvement and organisational outcomes. Proper implementation of AI in the working environment can increase employee happiness, but care must be given to address issues and maintain a healthy work environment in businesses [8]. Employees who are involved in what they do are more likely to give customer service quality above average, drive sales growth, and achieve business goals better than all competitors. In addition to that, organisations with high employee-engagement levels have lower absenteeism and turnover rates as well as a higher possibility of innovation and creativity. Therefore, concentrating on employee engagement is the key to success for enterprises with high growth levels and at the top of competitive markets.

### 3.4.2 AI-Powered Sentiment Analysis Tools for Measuring Employee Engagement

AI-based sentiment analysis tools provide a data-driven approach to evaluating and analysing employee engagement. These tools leverage NLP algorithms to analyse textual data in the form of employee surveys, feedback, and communication channels in order to gauge the overall sentiment and detect the underlying themes and patterns. Through the capture and analysis of unstructured data, sentiment analysis instruments provide operational findings of employee attitudes, levels of satisfaction, and areas for improvement.

There are multiple benefits in using AI-powered sentiment analysis tools for measuring employee engagement. First of all, they enable organisations to measure employee's mindsets in the present moment. This, in turn, allows managers to identify and deal with problems in a timely manner. Also, they address the issue of the attitudes and sentiments of employees that are expressed in different interactions like surveys, emails, and social media, creating a unified picture of what drives employee engagement. Eventually, they help organisations to track changes in employee sentiment as time passes, identify trends, and measure the effectiveness of engagement programmes.

### 3.4.3 Personalised Learning Platforms and Their Role in Enhancing Employee Engagement and Retention

AI-powered personalised learning platforms utilise ML algorithms and AI to provide workers with individualised learning experiences based on their unique learning preferences, learning styles, and skill deficiencies. AI provides more efficient training and development support for staff members. This can assist in identifying skill gaps and creating appropriate training initiatives [13]. These portals use employee data, for example, performance ratings, career aspirations, and learning history, to produce recommendations for relevant training materials, learning tracks, and development paths. These platforms enable personalised and adaptive learning where employee engagement, knowledge retention, and skill acquisition are enhanced (Figure 3.3).

The influence on employee engagement and retention through the deployment of personalised learning platforms is multilayered. Secondly, it enables employees to make their own destiny by taking hold of their educational and self-development processes, therefore leading to a sense of autonomy and pride. Lastly, these personalised learning tools address multiple learning difficulties as special programmes meet different learning styles and preferences. Additionally, it stimulates and enhances

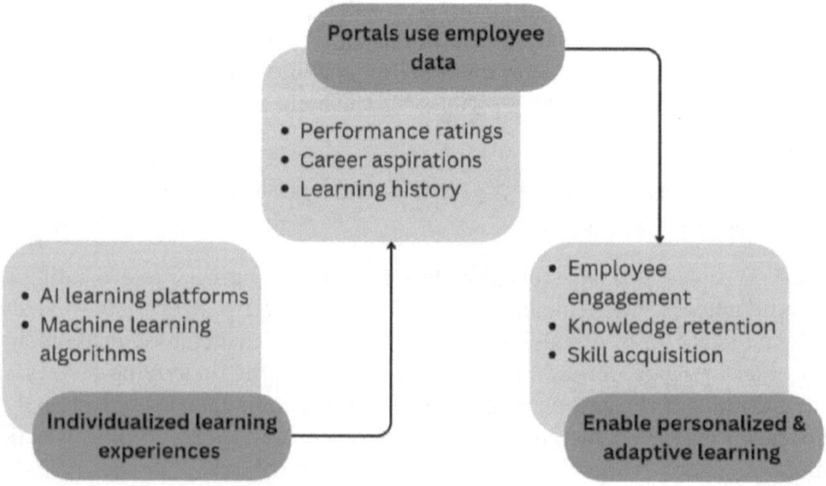


**FIGURE 3.3** Personalised learning through AI

**Source: Created by author**

the employees' knowledge with the newest practical and workplace-based skills, increasing job satisfaction and career growth opportunities.

In conclusion, the combination of AI sentiment analysis and individualised training technologies revolutionises the process of people engagement strategies. Through AI technologies, companies may now have access to more detailed information about employee feelings, be able to develop better learning, and thus improve employee engagement and productivity levels with higher retention.

## 3.5 ENHANCEMENT OF PERFORMANCE EVALUATION AND FEEDBACK MECHANISMS

Performance evaluation and feedback indeed contribute a great deal to the establishment of employee development, linking individual objectives with organisational goals, and bringing successful business accomplishments to the table. This further enhances the employee's happiness at workplace [19]. AI and further cutting-edge technology (published literatures) recommend improving HR practices and tactics to affect both individual and organisational performance in work environments [20]. In this section, we will investigate the drawbacks of the previous performance-evaluation system. After that, we will explain how AI-based performance appraisal systems address these issues and provide case studies showing how AI has been successfully implemented in performance evaluation and feedback.

### 3.5.1 Traditional Performance-Evaluation Processes and Their Shortcomings

The conventional performance appraisal process, which is oftentimes described by annual reviews and subjective assessments, carries significant disadvantages. One of their few main weaknesses is that they give more attention to past performance than to the future potential and development of their talents and abilities.

This stay-behind perspective would prevent the divergence from affected actions in real time and thus limit the timely adjustment and improvements. Following that, subjective evaluations that mainly depend on the manager's biases have always been used, and this can be a source of imbalance and unfairness in the evaluations. Additionally, the traditional performance-review process may be ambiguous and raise questions about accountability and leadership since employees lack awareness of their objectives, duties, and performance.

### 3.5.2 AI-Driven Performance Management Systems and Their Benefits

AI-driven systems for job performance evaluation provide an objective and data-oriented approach that helps to make up for the disadvantages of conventional methods of evaluating employee performance. Through automation, AI makes everyday jobs easier to complete successfully [21]. These types of systems are collections of AI algorithms and ML techniques that enable analysing performance data, finding patterns, and offering actionable recommendations (Figure 3.4).

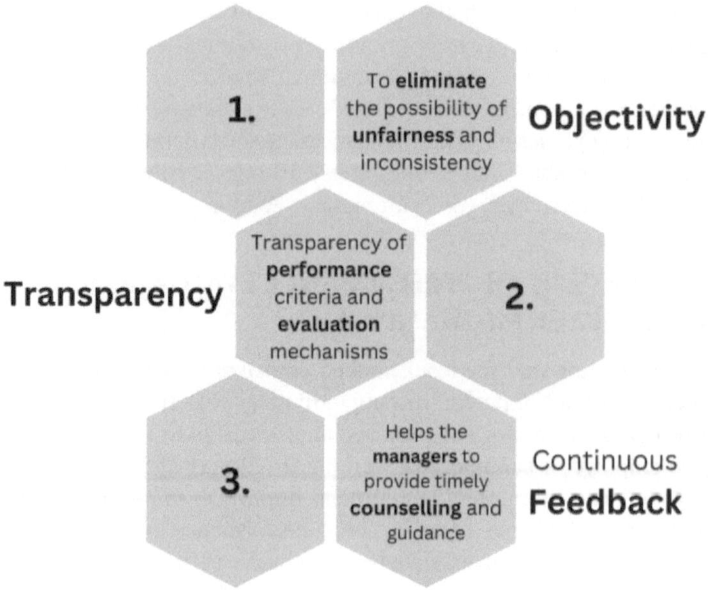


**FIGURE 3.4** Benefits of AI-driven performance management systems

**Source: Created by author**

**Objectivity:** AI algorithms have the ability to evaluate the performance of data equally, in cases where bias and subjective judgements are not involved. To eliminate the possibility of unfairness and inconsistency, performance reviews are determined by specific measures and algorithmic rules.

**Transparency:** AI-driven performance management systems enable transparency in performance criteria and evaluation mechanisms. This is done by using actual-time data and feedback for employees to evaluate their progress, dip into their strengths and weaknesses, and set very practical targets.

**Continuous Feedback:** AI-driven systems are also able to enable continuous feedback and suggestions for employees, which ultimately helps the managers provide timely counselling and guidance. Through comparisons of performance statistics and finding areas for improvement, these systems direct the organisations towards continuous improvement and development.

### 3.5.3 Case Studies: Implementation of AI in Performance Evaluation and Feedback

---

#### Case Study 1: General Electric[4]

The General Electric Company completely transformed its non-dynamic performance evaluation procedures by implementing an AI system for action performance management. One of the system's key features was the application of AI-based algorithms that could assess performance data, such as key performance indicators, peer reviews, employee engagement data, and others. Improved employee engagement and performance were the outcomes of the manager's ability to provide contextually aware feedback and one-on-one coaching to the staff member using the real-time AI-based system.

**Case Study 2: Adobe**[5]

Adobe introduced "Check-In," a new AI-influenced method, to replace the annual performance evaluation and ratings ten years ago. Prior to the implementation of "Check-In," employees expressed frustration with performance reviews. They considered the procedure to be complicated, time consuming, and burdensome. Key elements of the AI-powered performance management tool "Check-In" include the ability to explore new career options at Adobe, obtain real-time feedback, and put together a check-in dashboard. Over the past ten years, "Check-In" has strengthened Adobe's culture and helped the firm achieve its aim of giving employees the freedom to be who they are and the assistance to become who they want to be.

Cases that have been studied exhibit the life-changing power of AI-enabled performance management systems for performance assessment upgrades and feedback content. Through AI technologies, organisations will be able to solve the bottleneck problems of conventional techniques, create a feedback-oriented system, and enhance employee engagement, respectively.

## 3.6 AI OPENING UP NEW AVENUES FOR SUCCESSION PLANNING AND TALENT DEVELOPMENT

Talent development and succession planning are important in HRM, whose main goal is to make sure that every position has the right person who perfectly fits the company's strategic plan. Many people believe that integrating AI into HRM would change or replace employee responsibilities. Rather than replacing employee labour, AI improves administrative tasks and employee engagement [22]. AI in HRM can enhance team training, hiring, vacation request scheduling, and other daily HRM tasks. It can also improve organisational decision-making tasks [23]. We will start this part with a general description of talent development and succession planning, continue with explaining how AI-powered talent management systems operate in the context of identifying top talents and succession candidates, and end it with the role of AI-based modern platforms in individual skill development.

### 3.6.1 Succession Planning and Talent Development in HRM

Developing talent means inspiring and dynamising the capabilities and ambitions of employees to the benefit of the structure and its future. It involves a number of training programs like mentorship, counselling, career assistance, and so forth. Without particularity, succession planning aids in the identification and development of the top employees to assume key leadership posts in the organisation. Succession planning provides leadership continuity and stability in leadership roles by planning forward for the next generation of leaders. Without it, the organisation is vulnerable to risks associated with turnover and talent shortages.

### 3.6.2 AI-Powered Talent Management Platforms for Identifying High-Potential Employees and Succession Candidates

AI-powered systems embody principles of data and ML analytics, which enable the identification of high-potential employees along with succession leaders. The most notable feature of these platforms is their ability to examine a range of different data sources. They include performance evaluations, career trajectories, skill assessments, and employee reviews, among others. Such platforms allow for the identification of individuals who have the potential to rise in ranking into leadership positions (Figure 3.5). AI-customised training and onboarding that can be provided at the right moment [24]. Key features of AI-powered talent management platforms include the following.

**Predictive Analytics:** The AI algorithms that analyse historical data and the metrics consider past performance to forecast the future level of achievement. Through identifying trends and patterns, these sites will help to discover employees with the required intellectual tendencies and behavioural traits for leadership roles.

**Skills Mapping:** AI-based talent management systems check out the employees' competencies, skills, and deficits in their development life and then distinguish out any gaps or opportunities for growth. Through the process of matching abilities to organisational goals and future needs, HR managers are able to move their talent-development strategies closer to business strategy.

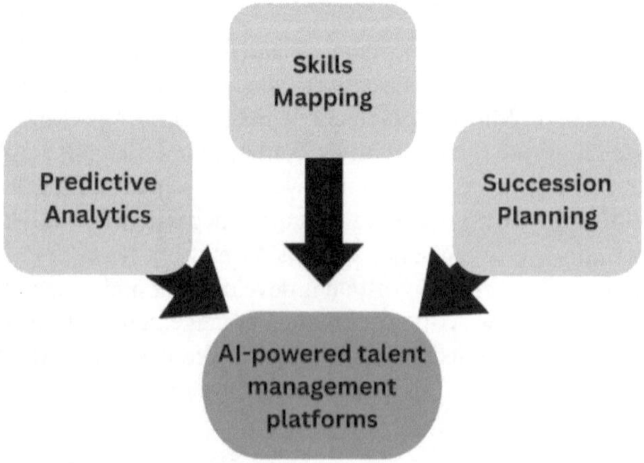


**FIGURE 3.5** AI-powered talent management platforms

**Source: Created by author**

**Succession Planning:** The AI-based talent management programme's strengths lie in Succession Planning, the process of finding and preparing prospective employees for key roles in the organisation based on their experiences, potential, and competencies. These systems build talent pipelines and succession candidates' training plans, which guarantee a smooth transition and limit disruptions in leadership continuance.

### 3.6.3 AI-Driven Learning and Development Platforms for Personalised Skill Development

AI-figured learning and development platforms make use of personalised and adaptive learning methods that allow employees to get within the scope of their learning resources. The AI-powered platforms use the analysis of the employee's learning styles, preferences, and the assessment of their performance data to match the recommended training content, the expertise building pathways, and the upskilling chances. Key benefits of AI-driven learning and development platforms include the following.

**Personalisation:** AI-based algorithms personalise learning experiences based on each individual employee's unique abilities, preferences, and knowledge gaps, thus engaging the individuals and promoting an increase in knowledge retention. These channels are equalled by the fact that they release only noteworthy and timely information, allowing employees to master the new skills and abilities autonomously.

**Adaptability:** AI-based learning and development platforms that work interactively with employees follow and take into account the employees' advancement, as well as their feedback; thereby, they can adjust the learning pathways and content recommendations according to those requirements. These technical platforms ensure that employees who aim to excel in their work get the tutoring, direction, and resources they need to attain their goals.

**Data-Driven Insights:** AI-driven platforms for learning and development in the HR department gather data on the employees' learning behaviours, among other things, such as their learning interests and performance. Through the observation of learning analytics and feedback data, HR experts are able to modify training programs, analyse performance, and determine return on investment.

In summary, AI-driven human capital management and training and development platforms carry exceptional capacity for talent growth and development planning in HRM. Companies can make use of AI methods to identify personnel with great potential, develop leaders of the future, and support workers of all levels so they have new skills and abilities. This procedure entails expanding the pool of applicants, boosting processing power, and employing AI to evaluate prospects more accurately [24]. Companies have the opportunity to obtain an edge and become top-notch in the era of digital technology.

## 3.7 RESPONSIBLE AND ETHICAL IMPLEMENTATION OF AI IN HRM

The expansion of AI in HRM makes a lot of sense because of the advantages that this will bring to innovative activities and the efficiency of operations. The increasing acceptance of AI and its capabilities has led to an increase in the focus on the ethical concepts and values that guide AI development and use [25]. Although there are some ethical questions that should be taken into account if this is to be implemented, the technology has a lot of advantages, such as fairness, transparency, and respect for employees' rights, thus giving more focus to other significant ethical issues. In this section, we proceed with the discussion focused on concerns about ethical aspects of AI implementation, the mitigation of bias, as well as transparency, privacy, and employee information security considerations. Moreover, we offer recommendations for responsible AI implementation in HRM.

### 3.7.1 Significance of Ethical Considerations in the Application of AI

Ethical considerations become absolutely crucial when it comes to AI implementation in HRM because potentially such AI-implementation-based decisions may affect the lives and incomes of people irreversibly. Fair and ethical use of AI implies following fairness, responsibility, transparency, and respect for human rights. Organisations need to deliberate on the ethical aspects of AI algorithms, data usage, decision processes, and the consequences produced by AI systems to ensure that they do not infringe on human ethical standards and values.

### 3.7.2 Ethical Considerations: Mitigating Bias, Ensuring Transparency, and Safeguarding Employee Privacy

**Mitigating Bias:** AI claims to mitigate bias throughout the procedure. Because ML algorithms incorporate human bias, AI has the potential to exacerbate this issue [21]. AI algorithms may be biased in both conscious and unconscious ways and could unintentionally propagate existing inequalities and discrimination. To minimise bias in AI-based HRM processes, organisations should have diversity representation in data collection, use a bias-detection algorithm that points out the biases present in training data, and audit AI systems on a regular basis for fairness and equity.

**Ensuring Transparency:** Transparency is key for nurturing trust and accountability in AI-powered HRM methods. Organisations need to provide clear explanations on how AI algorithms are used in the decision-making process, share data sources and model assumptions, and let employees know of the possible risks and limitations of AI technologies.

**Safeguarding Employee Privacy:** The privacy of employees is a basic right that has to be considered in AI-driven HRM processes. Organisations must put in place sound data-protection procedures, for instance, such as data anonymisation, encryption, and access controls, in order to protect sensitive employee data. Also, the organisations should have informed consent from the employees for data collection and usage and offer methods for people to view, edit, and delete their personal information.

### 3.7.3 Suggestions for Responsible AI Adoption in HRM

**Ethical Guidelines:** Establish ethical principles and guidelines for AI implementation in HRM. Outline the expectations for fairness, transparency, accountability, and respect for employee rights.

**Ethics Training:** Conduct ethics training and awareness programmes for the HR professionals and decision-makers responsible for implementing AI, making sure they know the ethical issues related to this technology and how to address them properly.

**Diversity and Inclusion:** Through diversity and inclusion in AI development teams, diverse perspectives are guaranteed in algorithm design, data collection, and the decision-making process.

**Bias Assessment:** Frequently conduct bias assessments and audits of AI mechanisms and decision-making processes to search for and remove biases that may affect employee outcomes.

**Transparency and Accountability:** Encourage transparency and accountability of AI-powered HRM practices through documentation of AI algorithms, data sources, decision-making processes, and commitments to assist those who might be subjected to any kind of unfair treatment or discrimination.

**Privacy Protection:** Introduce robust data security measures to protect employee data from unauthorised access, use, or disclosure, and ensure that you follow the proper data protection regulations of the countries that the company operates in, such as General Data Protection Regulation or California Consumer Privacy Act.

**Continuous Monitoring and Evaluation:** Keep on tracking and assessing the effects of AI-based HRM practices on employees' outcomes, engagement, and satisfaction; do the needle-based corrections if deemed necessary to rectify the unintended consequences and ethical issues (Figure 3.6).

Through facilitating these recommendations and putting ethics at the core of AI implementation, entities will be able to leverage AI to transform HRM while at the same time conforming to ethical guidelines and promoting trust, equity, and worker's rights.

## 3.8 FUTURE DIRECTIONS, RESEARCH AGENDA, AND CONCLUSION

People in low-income nations may view AI as providing a fresh opportunity to break the cycle of poverty, while people in advanced countries may fear losing their jobs owing to the technology [26]. Therefore, it becomes important that HR procedures need to include AI to become more efficient if they want to gain a competitive advantage in the context of escalating worldwide competition [27]. With the ongoing trend towards the implantation of AI in HRM strategies, we are certain that AI will keep changing and advancing the field in the future. The conclusion will saliently be a speculation on the future of AI in HRM, summarise the main ideas from the chapter, and underline the transformational impact of AI in the reshaping of HRM approaches that eventually enhance organisational success in the digital world.

### 3.8.1 Potential AI Applications in HR and Probable Future Development & Research Areas

The future of AI in HRM holds exciting possibilities, with several potential developments on the horizon.

**FIGURE 3.6** Responsible AI adoption in Human Resource Management

**Source: Created by author**

**Advanced Predictive Analytics:** As AI-powered predictive analytics evolve, organisations will be more capable of anticipating future workforce trends, spotting emerging talent needs, and developing proactive strategies that are adapted to them.

**Augmented Decision-Making:** AI-guided decision support systems will team up with human decision-making in HR, where the systems will analyse data and provide real-time insights to understand the labour market and optimise the organisation's performance.

**Ethical AI Governance:** There will be an increased focus on ethical AI governance in HRM, with organisations implementing robust frameworks and mechanisms to ensure transparency, fairness, and accountability in the decision-making processes driven by AI.

**Personalised Employee Experiences:** AI-based solutions will help companies develop targeted and individualised career experience packages that include accommodations for changing mental states, quality of life, and career objectives, thus resulting in high levels of engagement, satisfaction, and retention.

**Integration with Emerging Technologies:** AI will accompany trendy technologies like augmented reality (AR), virtual reality (VR), and blockchain to provide interactive and safe HRM solutions for vacuuming, training, assessment, and talent development.

There is a vast and potential area of research work in all the above potential applications and developments of AI that can be used in HR in the future. These AI-related fields, including AI-driven predictive analytics, AI-guided decision support systems, ethical AI governance in HRM, AI-based solutions, AR, and VR, could be studied by academics and researchers in the future.

### 3.8.2 Summary of Key Insights from the Chapter

One of the main points in this chapter is about the changes in human capital management due to the AI revolution in different aspects: recruitment, employee engagement, performance evaluation, talent development, and succession planning. Managers and leaders can significantly use their agency to promote the well-being of their workforce by influencing AI, technology adoption, and enhancing working conditions. [8].

#### 3.8.2.1 Key Insights

AI-driven resources and technologies allow employers to execute their HRM operations faster, more precisely, and effectively, with great potential. This results in the ability of the organisations to streamline operations, make better management decisions, and be more innovative.

It should be noted that the ethical issues of AI implementation are usually on the agenda of organisations that have to deal with problems like bias mitigation, privacy, and transparency issues, which gives a shortcut for using responsible AI techniques.

AI can be a life-changing factor in revolutionising the HRM functions and the organisation as a whole, but the necessity for ethics, law, and society's implications must be addressed properly.

### 3.8.3 Conclusion: Emphasising AI's Potential to Transform HRM Practices and Promote Organisational Success in the Digital Age

In conclusion, the disruptive force that has been brought to HRM practices by AI is impossible to underestimate. There are many disruptions occurring in the field of HR management, and as HR students, HR professionals, academics, and practitioners, we have good reasons to pay attention to these disruptions, revise regulations, and turn these disruptions into innovation. By thoroughly analysing various published manuscripts and previous studies using content analysis methodology, this chapter provides some valuable insights into how disruption turned into innovation for the HRM. Through AI technology, companies can enhance and negotiate recruitment procedures, improve employee involvement and efficiency assessment, enhance education and careers, and manage all resources for the company's performance and success in the age of digitalisation. Nevertheless, the potential of this development is huge; however, this capability can only be achieved by designing complementary strategies that enhance the ethical aspects of AI in addition to the transparency and accountability aspects. The incorporation of responsible AI applications and leveraging their transformative nature should be the way out for organisations as they strive to unlock the new opportunities that are required for growth, competition, and development in the globalised business community.

## NOTES

1 Source: Cambridge Dictionary; https://dictionary.cambridge.org/dictionary/english/disruption, extracted on March 10, 2024.
2 Source: https://www.linkedin.com/pulse/transformative-impact-ai-recruitment-processes-bizz-o-tech-2ninf/, extracted and revised on July 13, 2024.
3 Source: Ten HR Trends in the Age of Artificial Intelligence (forbes.com), extracted on March 12, 2024.
4 Source: GE's Real-Time Performance Development (hbr.org), extracted on March 13, 2024.
5 Source: How we inspire great performance at Adobe | Adobe Blog, extracted on March 13, 2024.

## REFERENCES

1. Minbaeva, Dana. "Disrupted HR?" *Human Resource Management Review* 31, no. 4 (2021): 100820. https://doi.org/10.1016/j.hrmr.2020.100820
2. Varma, Arup, Cedric Dawkins, and Kaushik Chaudhuri. "Artificial intelligence and people management: A critical assessment through the ethical lens." *Human Resource Management Review* 33, no. 1 (2023): 100923. https://doi.org/10.1016/j.hrmr.2022.100923
3. Prakash, C., R. Yadav, A. Singh, & K. Yadav. "Modeling the entrepreneurial intention among business students." *Pacific Business Review International* 14, no. 3 (2021): 77–90.
4. Bulut, Abdurrahim, and Özlem Batur Dinler. "The effect of industry 4.0 and artificial intelligence on human resource management." *Uluslararası Doğu Anadolu Fen Mühendislik ve Tasarım Dergisi* 5, no. 2 (2023): 143–166. https://doi.org/10.47898/ijeased.1306881
5. George, Ginu, and Mary Rani Thomas. "Integration of artificial intelligence in human resource." *International Journal of Innovative Technology and Exploring Engineering* 9, no. 2 (2019): 5069–5073. https://doi.org/10.35940/ijitee.L3364.129219
6. Sharma, Dipanker, Waleed Salehi, Bhawana Bhardwaj, Mohinder Chand, and Hasiba Salihy. "Dovetailing the human resource management with the cloud computing in the era of industry 4.0: A review." *Frontiers in Management and Business* 4, no. 2 (2023): 340–351. https://doi.org/10.25082/FMB.2023.02.004
7. Datta, Ashis, Md Ruhul Islam, Amit Kumar Mukherjee, and Debdatta Kandar. "Cloud computing: A solution to human resource management system." In *2012 International Conference on Radar, Communication and Computing (ICRCC)*, pp. 176–179. IEEE, 2012. https://doi.org/10.1109/ICRCC.2012.6450571
8. Chaudhary, Meena, Dipanker Sharma, Mohinder Chand Dhiman, and Ankita Chaudhary. "How human resource managers can utilize AI to promote employee well-being." In B. Bhardwaj, D. Sharma, & M. C. Dhiman (Eds.), *AI and Emotional Intelligence for Modern Business Management*, pp. 263–281. IGI Global, 2023. https://doi.org/10.4018/979-8-3693-0418-1.ch018
9. Karatop, Buket, Cemalettin Kubat, and Özer Uygun. "Talent management in manufacturing system using fuzzy logic approach." *Computers & Industrial Engineering* 86 (2015): 127–136. https://doi.org/10.1016/j.cie.2014.09.015

10. Bhardwaj, Bhawana, and Dipanker Sharma. "Migration of skilled professionals across the border: Brain drain or brain gain?" *European Management Journal* 41, no. 6 (2023): 1021–1033. https://doi.org/10.1016/j.emj.2022.12.011
11. Black, J. Stewart, and Patrick van Esch. "AI-enabled recruiting in the war for talent." *Business Horizons* 64, no. 4 (2021): 513–524. https://doi.org/10.1016/j.bushor.2021.02.015
12. Sposato, Martin. "Remote working in the time of covid-19: Developing a web-based community." *International Journal of Web Based Communities* 17, no. 1 (2021): 1–8. https://doi.org/10.1504/IJWBC.2021.112862
13. Budhwar, Pawan, Ashish Malik, MT Thedushika De Silva, and Praveena Thevisuthan. "Artificial intelligence–challenges and opportunities for international HRM: A review and research agenda." *The International Journal of Human Resource M*anagement 33, no. 6 (2022): 1065–1097. https://doi.org/10.1080/09585192.2022.2035161
14. Figueroa-Armijos, Maria, Brent B. Clark, and Serge P. da Motta Veiga. "Ethical perceptions of AI in hiring and organizational trust: The role of performance expectancy and social influence." *Journal of Business Ethics* 186, no. 1 (2023): 179–197. https://doi.org/10.1007/s10551-022-05166-2
15. Hu, Qirui. "Unilever's practice on AI-based recruitment." *Highlights in Business, Economics and M*anagement 16 (2023): 256–263. https://doi.org/10.54097/hbem.v16i.10565
16. Black, J. Stewart, and Patrick van Esch. "AI-enabled recruiting: What is it and how should a manager use it?" *Business Horizons* 63, no. 2 (2020): 215–226. https://doi.org/10.1016/j.bushor.2019.12.001
17. Dutta, Debolina, Sushanta Kumar Mishra, and Divya Tyagi. "Augmented employee voice and employee engagement using artificial intelligence-enabled chatbots: A field study." *The International Journal of Human Resource Management* 34, no. 12 (2023): 2451–2480. https://doi.org/10.1080/09585192.2022.2085525
18. Malik, Ashish, Pawan Budhwar, Hrishi Mohan, and Srikanth NR. "Employee experience–the missing link for engaging employees: Insights from an MNE's AI-based HR ecosystem." *Human Resource Management* 62, no. 1 (2023): 97–115. https://doi.org/10.1002/hrm.22133
19. Jaswal, Nidhi, Dipanker Sharma, Bhawana Bhardwaj, and Sascha Kraus. "Promoting well-being through happiness at work: A systematic literature review and future research agenda." *Management Decision* 62, no. 13 (2024): 332–369. https://doi.org/10.1108/MD-08-2023-1492
20. Vrontis, Demetris, Michael Christofi, Vijay Pereira, Shlomo Tarba, Anna Makrides, and Eleni Trichina. "Artificial intelligence, robotics, advanced technologies and human resource management: A systematic review." *The International Journal of Human Resource Management* 33, no. 6 (2021): 1237–1266. https://doi:10.1080/09585192.2020.1871398
21. Ore, Olajide, and Martin Sposato. "Opportunities and risks of artificial intelligence in recruitment and selection." *International Journal of Organizational Analysis* 30, no. 6 (2022): 1771–1782. https://doi.org/10.1108/IJOA-07-2020-2291
22. Islami, Khansa, and Dan Sopiah. "Artificial intelligence in human resources in the era of society 5.0." *International Journal of Research and Innovation in Social Science* 6, no. 11 (2022): 675–681. https://doi.org/10.47772/IJRISS.2022.61131
23. Matsa, Prasanna, and Kusuma Gullamajji. "To study impact of artificial intelligence on human resource management." *International Research Journal of Engineering and Technology* 6, no. 8 (2019): 1229–1238.
24. Johnson, Brad A.M., Jerrell D. Coggburn, and Jared J. Llorens. "Artificial intelligence and public human resource management: Questions for research and practice." *Public Personnel Management* 51, no. 4 (2022): 538–562. https://doi.org/10.1177/00910260221126498

25. Hermann, Erik. "Leveraging artificial intelligence in marketing for social good—An ethical perspective." *Journal of Business Ethics* 179, no. 1 (2022): 43–61. https://doi.org/10.1007/s10551-021-04843-y
26. Lohr, S. "From agriculture to art–the AI wave sweeps in." *New York Times* 19 (2018): 7.
27. Sharma, Anshima, Bhawana Bhardwaj, and Dipanker Sharma. "Artificial intelligence: Shifting the landscape of the human resources ecosystem." In B. Bhardwaj, D. Sharma, & M. C. Dhiman (Eds.), *Leveraging AI and Emotional Intelligence in Contemporary Business Organizations*, pp. 371–381. IGI Global, 2024. https://doi.org/10.4018/979-8-3693-1902-4.ch022

# 4 Data Quality Management Framework for Data Warehouse

*Rakesh Kr. Arora, Manoj Kr. Gupta, and Suraj Pal Singh*

## 4.1 INTRODUCTION

In today's world, huge volume of data is generated/gathered/captured through various software, social networking applications, sensors, etc. To handle, organize and store this voluminous data, there is a requirement of efficient storage technology that will be able to store, organize and handle this voluminous data efficiently and effectively [1–6]. Apart from this, the technology should support the analysis of this data to extract novel information, i.e. knowledge from it. Data warehousing has the capability to store, organize and handle huge volume of data. According to Inmon [5], "A data warehouse is a subject-oriented, integrated, non-volatile, and time-variant collection of data in support of management's decisions". In data warehouses, the data can also be stored using multi-dimensional structure that will be helpful to analyze the data based on multiple dimensions. Analysis of the data stored in data warehouses is performed using data mining and online analytical processing (OLAP) [7, 3, 6]. OLAP can be used to analyze the data for the known data analytics requirements, whereas data mining is used to extract the unknown, novel and useful patterns, i.e. knowledge. Therefore, OLAP is used to answer the question "What has happened?", whereas data mining is used to answer the questions "Why it had happened? What will happen in the future?" Data warehousing has proven its various applications in the field of education, e-governance, business, and medical science [1, 3].

The nature of information can be viewed as legitimately corresponding to the data esteem, which implies that the nature of information decides the genuine embodiment of its application that helps in business basic leadership process. If information put away in data warehouse isn't precise, solid and trustworthy, then more exertion will be devoted on manual exercises, revision and finally adjustment of the framework than on running the business. Therefore, an answer is required for effective data quality administration, which once enforced will guarantee higher quality of information [8–10].

DOI: 10.1201/9781032700953-5

## 4.2 DATA QUALITY MANAGEMENT

Management of data quality involves preparation of jobs, duties, strategies and systems concerning the procurement, upkeep and scattering of information. It is better to stress more on the database level and has recommended that if legitimate jobs are characterized for the clients who communicate with the framework, then a great deal of irregularities can be avoided [11–13].

Data quality management lifecycle must be plainly characterized utilizing nonstop just as iterative structures. Good data quality empowers decision support systems and improves business reaction. Robotized purging and checking of data guarantees fast improvement in data quality and furthermore averts blunders by human. The use of statistical techniques guarantees consistent observation of data quality and gives extra controls to follow information-purifying endeavors.

## 4.3 WHY DATA QUALITY MANAGEMENT?

The data quality is a typical issue that most associations face while dealing with their data warehouse. A data warehouse can't provide an attractive return except if the data within the framework is exact, dependable and sound. Hence to guarantee that the data kept up in data warehouse is significant from business point, an efficient and effective data quality management framework is required. To provide the data with precision and unwavering quality, the quality of data-confirmation method should be incorporated in data analysis, cleansing, standardization and validation [11–13, 9, 14].

Improvement of data is a basic requirement with respect to business announcing. Data loss during relocation process results in loss of information that severely influences decision-making process. Organizations can neither bear to disregard cost sparing because of decrease in repetition of information and disposal of redundant data nor the expense of wrong choices dependent on inaccurate information.

The answer for organizations for data quality affirmation is to guarantee solid implementation of business rules and friendly environment to implement application, which enables operators to make precise queries.

## 4.4 BARRIERS TO DATA QUALITY MANAGEMENT

Productive management of data is a basic necessity to ensure data quality. Following are the hindrances to data quality management process [8, 9, 15–19]:

- Different data-entry operators making numerous master records in the framework, which results in the duplicity of data.
- Absence of standard conventions in business rules results in inconsistency.
- Missing information components, for example, shipping date and request date, will make tracking difficult.
- Absence of relationship links among dependent entities results in vagueness and repetition.
- Forcing data quality measures without required data.
- Unstandardized database mapping makes irregularity in the database.

## 4.5 CONTINUOUS AND ITERATIVE DATA QUALITY MANAGEMENT FRAMEWORK

The continuous and iterative approach helps in efficient and effective data quality control. Based on this approach, the system persistently monitors the nature of information and gives vital contributions to the frameworks to take convenient activities to control the quality. As quality improves iteratively over time, so what is required is an iterative system that consistently focuses to improve the information quality. As presented in Figure 4.1, the continuous and iterative data quality management framework for data warehouse comprises four iterative stages.

### 4.5.1 Data Analysis

Data analysis or investigation is the way toward looking at and concentrating the data with a target to check exactness and consistency of information dependent on the business rationale and connections. This incorporates contrasting the database diagram and the business rationale and relationship existing among focused substances. Data is assessed for mistakes, inconsistencies, duplications, missing values and deficiencies. This stage comprises following three iterative steps:

- **Detection:** This step is the initial move towards ensuring the data quality. Clever calculations and devices are used to constantly monitor and inform about the nature of the data and display various inconsistencies if detected.

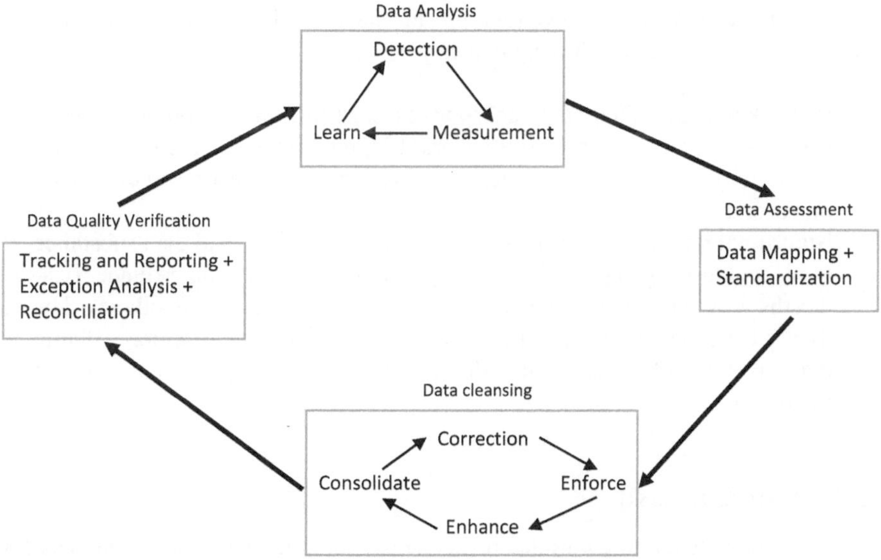


**FIGURE 4.1** Continuous and iterative data quality management framework for data warehouse

- **Measurement:** This step includes estimation of the information quality and subsequent mistakes. It focuses to distinguish peculiarities by legitimate observing procedures and key execution markers. Data errors can be grouped as per explicit procedure-dependent characterization. When relationship is expected, Type I error happens; however, in reality the relationship does not exist. This prompts extra approval and false cautions. When no relationship is expected, Type II errors happen; however, in reality the relationship exists. Recognizing Type II errors is troublesome, as they require top to bottom information investigation. Characterizing certainty breaking points of the information mistakes helps in organizing the exertion that should be spent to bring the information quality within as far as possible.
- **Learn:** This step includes looking into the procedure intermittently and upgrading the capacity to distinguish abnormalities by improving business forms. This progression guarantees that exertion requirement to maintain information quality diminishes after some time.

### 4.5.2 Data Assessment

The activity to ensure data quality requires well-characterized procedures that need to be pursued for boosting the quality. This process for quality control includes evaluation of various issues regarding data quality, cleansing of data, expansion of data and determining and revealing of various irregularities in data. The assessment phase of data comprises examination of analysis report and concluding the extent of quality of data reviews. It is a significant stage for management of data as needs of rectification endeavors are resolved within this stage. In this stage, data assessment is accomplished using the following two steps:

- **Data Mapping:** This step incorporates start-to-finish mapping between source framework and goal framework. Mapping improves greater perceivability to the cleaning necessities. Purging reports are made and organized on substances.
- **Standardization:** Standardization is imperative to overhaul data substance to meet industry gauges and business guidelines. Data should be made trustworthy crosswise over frameworks, which thus diminishes redundancy. The data standardization makes it convenient to compare two or more different datasets, which belong to same domain, that are having different range of the data values.

### 4.5.3 Data Cleansing

Data cleansing is taken to guarantee respectability of data and to set up the data for explicit movement requirements. Number of tools and smart calculations tweaked to business situations are executed to distinguish the infringement of the standards. This stage satisfies the following accompanying goals:

- Confirm that the data should meet the necessities of forthcoming framework.
- To minimize movement-associated mistakes, therefore lessening manual information section exertion.

Data cleansing stage comprises the following four iterative steps:

- **Correction:** This includes taking important activities to address the information peculiarities. While expository calculations are utilized to auto-correct a portion of the sections, client mediation is called for when there are number of choices for the move that can be made. The following four actions can be taken while transferring data into the data warehouse.
- **Enforce:** Data warehouses can be utilized for implementing referential trustworthiness and requirements. Extra business imperatives can be authorized for utilizing pre-characterized triggers on business databases. Some of the imperatives that can be authorized include:

    - Referential limitations to confirm whether key is available in referential super table before exchange record can be manufactured.
    - New imperatives that ensure copies inside a substance are maintained a strategic distance from.
    - Default imperatives to connect default esteems without client passage in business databases.

- **Enhance:** This step includes augmentation of data by improving the data values utilizing inward and outside information sources. Data can likewise be upgraded by sourcing data from frameworks inside the association by coordinating the necessary substance over the applications. For models, different clients in database are connected together with data in outer database, which brought about progressively exact revealing profitability for the client.
- **Consolidate:** This step includes minimizing redundant or superfluous information in the data warehouse. This guarantees that transactions are applied to accurate dimensions making rolling up simpler.

### 4.5.4 Data Quality Verification

This step centers around extra endeavors to check the nature of data by conducting live test. Cyclic validation approach tests databases and analysis is done to preconfirm about the nature of the data. The essential information regarding the quality of data is then provided to administration. In this stage, verification of data quality is accomplished using the following three steps:

- **Tracking and Reporting:** This activity is essential to comprehend preparation of data. It empowers clients to determine the effectiveness of the correction. It gives the privilege to prioritize based on seriousness of data quality issues. Tracking recognizes the most widely recognized reasons for errors and helps to investigate the conceivable outcomes of mechanizing amendment systems.

- **Exception Analysis:** Data validation assumes a significant step in data movement. It demonstrates to business network that the final product of data purging is being done and furthermore gives perceivability into various types of errors along with their causes, and restorative activity that should be considered. In-depth examination of blunders experienced, amid relocation, measurable outline of failed records, following of accomplishment level of every change run, and furthermore proactively distinguishes naive territories of purging to accomplish better transformation rates.
- **Reconciliation:** In this step, comparison and analysis of data from both source and target framework is done. Efforts are made to bring both source and target information to a typical system for examination. Budgetary compromise like explicit matching of income and expense information from source and goal frameworks to guarantee that income stream stays steady.

## 4.6 CONCLUSION AND FUTURE SCOPE

DQM (Data Quality Management) is essential for the accuracy, consistency and reliability of the data in the DW (Data Warehouse). However, there are several limitations and challenges that an organization may face when implementing a data quality management framework. Some common limitations include the following: Data complexity: Data in a DW often comes from different sources with varying formats, structures and quality levels. It can be difficult to integrate and manage data from diverse sources. Data volume: As data volume increases, it becomes more difficult to maintain data quality. Scalability issues can occur when dealing with large data volumes, which can lead to delays and increased processing time. Data governance: Lack of standards can impede the effectiveness of a data quality framework. Without uniform standards, it can be difficult to consistently enforce data quality rules. Data transformation: ETL (Extract, Transform and Load) processes are critical for preparing data for storing in a DW. Errors or issues during ETL processes can lead to data quality issues. Monitoring and validating data transformations is important [20, 21].

Implementing data quality management program isn't a simple assignment; however, the prizes are gigantic. Conveying a restrained way to deal with overseeing data improves the profitability of its data specialists and to more readily serve its clients.

Solid systems and procedures are basis for overseeing the most important corporate resource, that is, data for maintaining quality of data. Smart calculations help in the identification of errors hereby limiting the risks. With the help of additional validation techniques, for example, exception analysis and reconciliation guarantee better success rates in relocation activities.

The proposed framework is scalable and can be extended for other business requirements for migration, business intelligence warehouses, and so on. Maintenance costs can be greatly reduced by implementing proposed framework. Effective handling of the challenges related to quality of data can be done by using the proposed framework.

## REFERENCES


1. Arora, R.K. and Gupta, M.K. (2017), e-Governance Using Data Warehousing and Data Mining. *International Journal of Computer Applications*, Vol. 169, No. 8, pp. 28–31.
2. Badal, D., Arora, R.K. and Gupta, M.K. (2011), Data Warehouse Model for University. In *Proceedings: International Conference on Transformation & Survival of Business Organizations: Challenges and Opportunities,* Prestige Institute of Management, Gwalior, MP (ISBN: 978–81–8487-225-5, pp. 200–203).
3. Chandra, P. and Gupta, M.K. (2017), Comprehensive Survey on Data Warehousing Research. *International Journal of Information Technology*, Vol. 10, No. 2, pp. 217–224. https://doi.org/10.1007/s41870-017-0067-y.
4. Gosain A., Heena. (2015), Literature Review of Data Model Quality Metrics of Data Warehouse. In *Procedia, Computer Science,* Elsevier, *International Conference on Intelligent Computing, Communication and Convergence* (ICCC-2014)
5. Inmon, W.H. (2005), *Building the Data Warehouse*, 5th edn. New York: Wiley.
6. Ponniah, P. (2001), *Data Warehousing Fundamentals.* New York: Wiley.
7. Berson, A. (2017), *Data Warehousing, Data Mining & OLAP.* The McGraw Hill.
8. IDRBT (2014), *Data Quality Framework for Indian Banking Sector.* An IDRBT Publication.
9. Kimpball, R. (2007), *An Architecture for Data Quality.* A Kimpball Group White Paper.
10. Prakash, C., Yadav, R., Dangi, A. and Singh, A. (2023), Factor Analysis Approach to Study Mobile Applications' Characteristics and Consumers' Attitudes. In *Computational Intelligence for Engineering and Management Applications: Select Proceedings of CIEMA 2022* (pp. 783–798). Singapore: Springer Nature Singapore.
11. Benkhaled, H. and Berrabah, D. (2019), Data Quality Management for Data Warehouse Systems: State of the Art. In *Proceedings* 3rd Edition *of the National Study Day on Research on Computer Sciences (JERI 2019),* At Saida, Algeria.
12. Geiger, J.G. Data Quality Management the Most Critical Initiative You Can Implement, *Data Warehousing, Management and Quality*, SUGI 29, Paper 098-29 https://support.sas.com/resources/papers/proceedings/proceedings/sugi29/098-29.pdf
13. Helfert, M. (2001), Managing and Measuring Data Quality in Data Warehousing. In *Proceedings: World Multiconference on Systemics, Cybernetics and Informatics*, pp. 55–65.
14. Shanks, G. and Darke, P. (1998), Understanding Data Quality in Data Warehousing: A semiotic approach. In *Proceedings of the MIT Conference on Information Quality* (pp. 247–264). Boston, MA, Cambridge: Massachusetts Institute of Technology.
15. Nemani, R.R. and Konda, R. (2009), *A Framework for Data Quality in Data Warehousing*, UNISCON 2009, LNBIP 20, pp. 292–297. Berlin Heidelberg: Springer-Verlag.
16. Pipino, L.L., Lee, Y.W. and Wang, R.Y. (2002), Data Quality Assessment. *Communications of the ACM - Supporting Community and Building Social Capital, ACM New York*, Vol. 45, No. 4, pp. 211–218. https://doi.org/10.1145/505248.506010
17. Singh, S.P. and Solanki, S. (2019), A Movie Recommender System Using Modified Cuckoo Search. In Sridhar, V., Padma, M. and Rao, K. (eds) *Emerging Research in Electronics, Computer Science and Technology. Lecture Notes in Electrical Engineering*, vol. 545. Springer.
18. Singh, S.P. and Solanki, S. (2019), Recommender System Survey: Clustering to Nature Inspired Algorithm. In Krishna, C., Dutta, M. and Kumar, R. (eds) *Proceedings of 2nd International Conference on Communication, Computing and Networking. Lecture Notes in Networks and Systems*, vol. 46. Singapore: Springer.
19. Agrawal, S., Singh, S.P. and Sharma, J.K. (2018), Predicting Results of Indian Premier League T-20 Matches using Machine Learning. In *2018 8th International Conference*

*on Communication Systems and Network Technologies (CSNT)* (pp. 67–71). Bhopal, India. https://doi.org/10.1109/CSNT.2018.8820235.

20. Al-Okaily, A., Al-Okaily, M., Teoh, A.P. and Al-Debei, M.M. (2023), An Empirical Study on Data Warehouse Systems Effectiveness: The Case of Jordanian Banks in the Business Intelligence Era. *EuroMed Journal of Business*, Vol. 18, No. 4, pp. 489–510.
21. Mansouri, Taha, Moghadam, Mohammad Reza Sadeghi, Monshizadeh, Fatemeh, Zareravasan, Ahad. (March 2023), IoT Data Quality Issues and Potential Solutions: A Literature Review. *The Computer Journal*, Vol. 66, No. 3, pp. 615–625.

# 5 An Empirical Investigation of Educators' Professional Expertise to Incorporate Artificial Intelligence (AI)-Based Tools into the Classroom in a Responsible and Ethical Manner

## *Approach Toward Intelligent-TPACK*

*Preeti Chawla and Meenakshi Bisla*

## 5.1 INTRODUCTION

According to certain academics and industry professionals, automation technologies like artificial intelligence (AI) will significantly alter the labor market [3, 9]. Some have voiced apprehension regarding the widespread use of AI, highlighting the possibility that it could spiral out of control and cause havoc in society [25]. Computerization and automation will undoubtedly impact how people work since AI will create new occupations or change existing ones. Seventy-five million to 375 million individuals, or 3–14% of the global workforce, may need to change careers or improve their skills by 2030, according to a 2017 McKinsey Global Institute report, depending on how quickly AI is adopted [52]. Furthermore, everyone in the workforce will have to adjust to operating ever-stronger machinery; thus, people must be suitably trained to fulfill these and related future employment demands. Employees

DOI: 10.1201/9781032700953-6

will need to redefine expectations for work and acquire in-demand skills to advance their careers and stay relevant.

The potential of AI to support learning in a variety of contexts has led to an increase in the field of education [39]. AI in education (AIEd) has many uses, such as intelligent tutors for content delivery, feedback provision, and progress monitoring. AIEd has also been shown to be a tool for theoretical innovations, technological breakthroughs, and positive pedagogical effects [58]. AIEd's affordances are widely recognized. AI can be utilized to increase knowledge-gap awareness and offer specialized help, allowing educators to deliver individualized and adaptable training with more effectiveness and efficiency [29]. Additionally, AI offers algorithm-based judgments that provide efficient, real-time evaluation of sophisticated knowledge and skills [11, 20]. AI-powered educational systems can also be used to assess student participation and classroom dynamics, which helps to detect at-risk kids in real time and allow for prompt intervention [77].

The instructional potential of AI has been actively pursued by researchers and practitioners, leading to a notable surge in scholarly production on AIEd [33]. Scientific literature can be examined using research motive identification, scientific collaboration evaluation, and research topic detection. It is useful for fully comprehending the background and current state of a field [12]. It seems appropriate to synthesize the existing literature for a condensed perspective, given the explosive growth of AIEd research.

AIEd literature was reviewed by many researchers from four perspectives: technology-assisted assessment, innovative instructional strategies, personalized instructional materials, and communication between instructors and learners. AIEd's strengths and opportunities were examined by Roll et al. [47, 58], who discovered that there was a revolutionary process regarding the adoption of AI technologies in students' daily lives and community activities, as well as connections with educators adopted by AI expertise [10]. In their systematic assessment of AI in higher education, Zawacki-Richter et al. [79] found applications of AI in prediction and profiling, as well as reviews of AIEd based on quantitative approaches. To identify gaps in AIEd applications and theory, Guan [29] examined 400 publications on AI and deep learning (DL) in education. Tang [70] carried out a thorough assessment of publications about the use of AI in e-learning to predict future research directions. The analysis showed that Bayesian-network-based student characteristic prediction and AI-based personalized learning scenarios were widely used. Nevertheless, most of these evaluations have used qualitative approaches, analyzed a small number of studies, and discussed individual findings; as a result, they fall short of providing a comprehensive picture of the field, especially when it comes to study subjects and topic evolution [21, 80]. However, this manual coding and synthesizing method of analyzing a publication's entire contents takes a lot of time and effort, and as the amount of published material grows at a rapid pace, it becomes outdated. Researchers and educators can use the inferred knowledge to better plan projects and choose research topics since it provides a better grasp of the advancement of applicable technologies, historical and current research progress, and forces behind new ideas.

In a nutshell, AI is the area of computer science that studies how intelligent conduct is simulated in computers and how these machines might imitate—and hopefully improve upon—human behavior. In addition to its dominance in science, engineering, and technology, AI is also prevalent in education thanks to machine learning systems and algorithm development. For example, AI has many algorithmic applications in education, including automated assessment systems to help educators assess students' knowledge, facial recognition systems to reveal behavioral patterns of students, and personalized learning platforms to encourage student learning.

After that, we provide a basic definition of AI and discuss how machine learning and algorithmic systems are used in these fields. We recognize that definitions of AI are complex, multidimensional, and contested in the literature, and as American scholars working in the field of education and educational technology, we only offer a brief definition below—a risky oversimplification given the nature of the work. A thorough discussion of these complexities is, however, outside the purview of this paper. Second, we go into greater depth about five AI-related educational applications and highlight the advantages they may have for both educators and students.

## 5.2 LITERATURE REVIEW

### 5.2.1 Definitions and Relevance of AI

With the development of AI, efforts to build intelligent machines that mimic human conduct have intensified. As a result of recent developments in computer science, there are many different definitions and justifications for what constitutes artificial AI systems. This definition emphasizes how human behavior and consciousness are mimicked.

Additionally, the definition of AI is given as follows: "the combination of cognitive automation, machine learning, reasoning, natural language processing, hypothesis generation and analysis, and intentional algorithm mutation producing insights and analytics at or above human capability".

Social media is being more and more incorporated into K-12 education, putting learners' activities under the control of sophisticated algorithmic systems. The term "K-12 education" in this context refers to the education that American kids get from kindergarten (K), which corresponds to primary and secondary schooling, through the 12th grade, which is equivalent to ages 17–18. Kindergarten through grade 12 is referred to as K-12 education in the present Indian educational system. Core education is taught during the first ten years of the system, and then pupils are prepared for two academic years of higher education in various subjects. These AI technologies can improve K-12 educational systems' functionality and aid in educators' and students' social and cognitive growth.

### 5.2.2 Theories of AI in Education

First off, there are just a few educational ideas present in the notable research, and they are not frequently accepted. This result was like that of the Zawacki-Richter et al. [79]

research. First, based on how they acquire and process knowledge, pupils are categorized using a learning-style model [26]. A unique Bayesian model for identifying learners' learning styles was described by García et al. [27]. The second aspect of situated learning theory is the influence of children's social environment on their growth. Thomaz and Breazeal [71] promoted a novel theory that saw machine learning problems as the interaction between humans and machines, which was influenced by the situated learning theory.

The idea holds that learning is a process of engaging in communities of practice that have similar norms, beliefs, and goals rather than a process of gaining abstract knowledge. Additionally, learning comes from engaging with the world and the people in it rather than being a distinct activity from doing. Thus, the idea of situated learning contends that education must be integrated into significant and applicable settings that mirror the actual circumstances in which the acquired information and abilities are utilized.

A key component of practice theory and information literacy is the idea of situated learning, which emphasizes learning as a socially situated activity in which "agent, activity, and the world mutually constitute each other" rather than as an internalized and individualistic practice [47].

### 5.2.3 Instructor/Educator Expertise for AI-Driven Learning

The opportunities demonstrate how AI has also significantly changed how instructional pedagogy is implemented [49]. Educators need to be aware of the pedagogical benefits of AI-based tools to fully take use of the prospects presented by AIEd [76]. When educators possess the necessary pedagogical understanding to make use of AI-based tools, they can effectively implement AI technology into their lessons [6]. For example, educators employ AI-based technologies more to promote learner motivation and engagement when they realize their benefits [40, 74]. In a similar vein, educators who possess a greater understanding of AI are better equipped to choose relevant AI-based teaching tools [24]. Therefore, understanding AI allows educators to deploy AI-based solutions for timely feedback and personalized learning [57]. It is crucial to comprehend their level of expertise when it comes to using AI-based solutions into the classroom. In actuality, the effective educational integration of any technology depends heavily on the role that technological and pedagogical knowledge (PK) plays [53]. Bozkurt et al. [4] also evaluated AIEd papers from 1970 to 2018 with social network analysis. After conducting a bibliometric analysis of 132 AIEd articles between 2007 and 2017, Hinojo-Lucena et al. [33] concluded that there was a global interest in AIEd and that the period under consideration was still in the early stages of the field's publishing history.

Future educator replacement by AI is not anticipated [37]. This is because educator–student interaction is essential to both the advancement of learning and the personal growth of each student [13]. However, because of their quick growth, AI and its subfields will surround learning and teaching environments [56]. Therefore, to implement AI-based education, educators' professional expertise will be transformed by AI [30, 31]. According to this perspective, the ability to use AI-based systems both pedagogically and technologically is essential for those in the teaching profession. The knowledge that educators need to integrate AI-based tools into their lessons could be explained by the technical, pedagogical, and content knowledge (TPACK) framework. According to Mishra and Koehler [53], TPACK is the professional knowledge that instructors possess to use technology for teaching efficiently. According to Mishra et al. [54] and Valtonen et al. [73], TPACK is seen as a flexible framework for a range of pedagogical approaches and technical tools. We assume that the TPACK framework will offer a strong foundation for an improved application of educators' skills for AI-based teaching when it is linked with AI's technological and pedagogical contributions. AI-based technologies offer educational benefits, but they also raise ethical questions. AI-based technologies have the potential to produce systematic and recurring errors in decision-making, as noted by Saini [61] and Shin [65]. These mistakes violate the inclusivity of education by leading to discrimination against pupils of different racial and cultural backgrounds [18, 20]. Certain AI-based language learning programs may be unable to distinguish between various gender voices [1, 2]. Like this, fairness issues might occasionally arise from automated scoring system outputs. Furthermore, educators may find it difficult to comprehend the reasoning behind the choices made by AI-based solutions. This could be because AI-based decisions are made mysteriously [5, 22]. Uncertainty on the part of the appropriate AI software developers may also give rise to ethical concerns about AI-based products [64]. Thus, it is a contentious matter when educators or learners are not given adequate knowledge about the accountable company for AI technology. All parties involved in education must think about and assess AI-based decisions due to ethical concerns [34]. Therefore, we contend that educators need to be knowledgeable about how to interpret, support, and assess data generated by AI-based systems. To develop an inclusive future generation, educators must possess this understanding of ethical assessments [35, 49]. Effective AI integration depends in large part on instructors' ethical judgments in addition to their technological and pedagogical expertise.

### 5.2.4 The Framework for TPACK

The foundation of the TPACK system is the pedagogical content knowledge (PCK) that Shulman [66, 67] proposed. According to Shulman [66], PCK is the comprehension of how specific aspects of the subject matter are organized, modified, and enhanced to improve student learning. As stated in various ways, PK and content knowledge (CK) are joined to generate PCK, which is the set of suitable techniques and approaches for instructing a certain subject. Technological knowledge (TK) is to gain a deeper understanding of educators' proficiency in technology integration. TPACK, technical content knowledge (TCK), and technical pedagogical knowledge (TPK) were the outcomes of the extension process. TPACK is a comprehensive and adaptable framework that has been embraced by several academics for various technical tools and educational approaches because of its flexibility [42, 45, 53, 79]. In the context of TPACK, for example, Lee and Tsai [48] determined knowledge to utilize web technologies. Furthermore,

Valtonen et al. [73] used TPACK to teach teamwork and critical thinking, two 21st-century abilities. The knowledge and abilities needed to use AI-based technologies are the focus of the current study. As a result, we concentrated on the knowledge components of TK, specifically TK, TCK, TPK, and TPACK. TK is defined as having the ability to use technology and an awareness of its benefits and drawbacks [54]. A desire to keep up with evolving technology is also covered by TK. Higher TK educators can use technology in both their work and personal lives. Moreover, they could quickly understand how much technology may help or impede completing a task [54]. The next definition of TCK is the comprehension of the relationships and limitations that exist between technology and CK. TCK comprises information regarding the technology used in the subject matter (geography and mathematics, for example) [41, 54]. Additionally, TCK considers how a certain technology might advance instructors' subject-specific expertise [53]. For example, a science instructor can use a chatbot to look for current topic content. Skill-enhancement programs can also be offered virtually through online modes that include e-coaching. TPK, on the other hand, is the understanding of the nature of technology-assisted teaching and learning. TPK includes information about the benefits and limitations of various technologies for teaching pedagogies [54]. Finally, the knowledge and abilities of the other essential components are integrated to make TPACK [8]. The idea of employing an appropriate technology tool to apply successful pedagogical practices when teaching a certain subject is known as TPACK. Or, to put it another way, educators who possess a higher level of TPACK competency can execute content-specific educational approaches using technology [54].

The following explanations of the Intelligent-TPACK elements are as follows:

- Intelligent-TK focuses on teaching people how to utilize and communicate with AI-based technologies and their core functions. This element seeks to gauge the degree of instructors' familiarity with the functionalities of AI-based technologies.
- Intelligent-TPK encompasses the pedagogical acquaintances of AI-powered tools, like learning monitoring and quick, tailored feedback for students. Additionally, instructors' comprehension of alerting (or notification) and their ability to analyze messages from AI-based systems are measured by Intelligent-TPK.
- The understanding of AI-based technologies unique to a certain sector is the emphasis of Intelligent-TCK. It gauges how much use educators make of AI-powered resources to keep their subject-matter expertise current. This component addresses educators' comprehension of technologies that are more suited to address subject-matter learning in their area of expertise.
- The core knowledge domain is thought to be intelligent TPACK. It evaluates the capacity of educators to choose and implement appropriate AI tools for implementing a teaching approach to satisfy learning objectives in a particular topic.
- Ethics gauges the instructor's opinion on the choice of AI-based resources. The evaluation is predicated on openness, equity, responsibility, and inclusivity.

## 5.3 DEVELOPING OF HYPOTHESIS

Educators need to be familiar with the fundamental technological operations of AI-based technologies to interact with them [30, 63]. Technical proficiency in using or interacting with AI-based technologies is necessary for successful AI-based training [49]. Successful technology integration in the classroom is favorably correlated with the technical proficiency of the educators [22, 45]. According to Kabakci Koehler and Mishra [44] and Schmid et al. [62], technological knowledge is seen to be a reliable indication of one's comprehension of the educational potential of relevant technology. Furthermore, instructors' awareness of pedagogical affordances increases with their level of technological proficiency [7, 54, 60]. Educators proficient in intelligent tutoring systems, for example, are inclined to use intelligent tools more appropriately tailored to their profession. Additionally, educators may use their technological expertise to assess each choice made using a specific technology tool in an ethical manner [1, 2, 17].

Consequently, we postulate the following for testing the conceptual model (Figure 5.1):

- H1. Intelligent-TPK benefits from Intelligent-TK: Educators' usage of AI-based essay scoring programs has been found to be positively impacted by their recognition of the advantages of these programs [74].
- H2. Intelligent-TCK benefits from Intelligent-TK: In a similar vein, [15] empirical research show that instructors' comprehension of chatbot tools increased the likelihood that they will be used. Increased usage of AI-based technologies may advance educators' expertise in integrating these resources into their teaching. Additionally, in the context of teaching English as a foreign language, Du and Gao [23] carried out research with instructors.

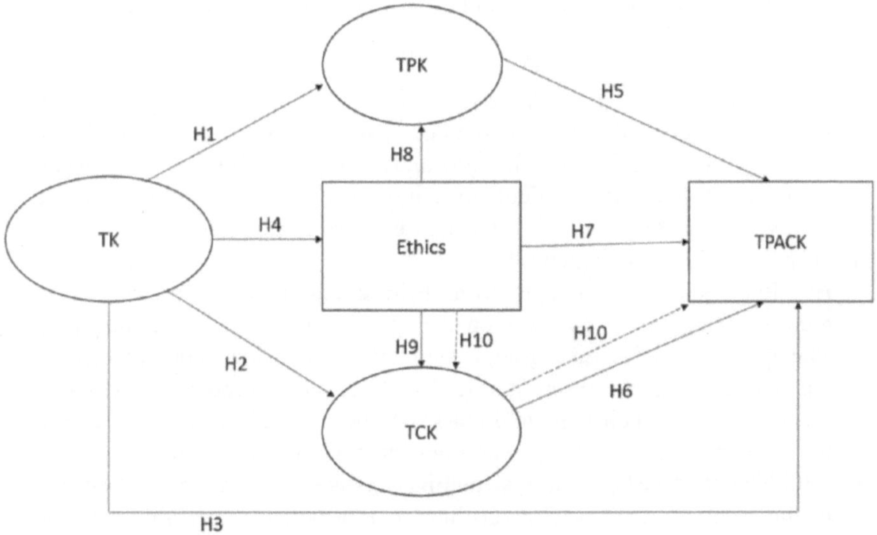


**FIGURE 5.1** Conceptual framework

- H3. Ethics benefits from Intelligent-TK: Educators get familiar with AI-based technologies if they use them to search for material unique to their profession [49]. Educators may also learn how to utilize these applications in their profession as they advise their colleagues about AI-based solutions that are particular to their field [69]. For example, if a STEM field instructor understands how AI-based technologies will alter topic understanding, they will be better able to incorporate AI-based tools into the classroom.
- H4. Intelligent-TPACK benefits from Intelligent-TK & H5. Intelligent-TPK benefits from Intelligent-TPACK: According to earlier studies, instructors' pedagogical usage of AI-based technologies is strongly correlated with their ethical assessments of AI-based choices. For example, Kabakci Yurdakul and Çoklar [42] discussed ethical competencies as TPACK core knowledge. Kabakci Yurdakul and Çoklar [42] and Yurdakul et al. [78] discovered that preservice educators' technological integration expertise and ethical competencies had a favorable correlation in another TPACK-based study. A better assessment of AI-based tools' judgments may come from educators' interactions with these tools [1, 2].
- H6. Intelligent-TPACK benefits from Intelligent-TCK: The pedagogical application of AI-based technologies in instruction may precede enhanced ethical assessment [55]. For example, an educator who is adept at evaluating the rationale behind decisions made by AI-based tools can make efficient use of AI-based tools in the classroom.
- H7. Intelligent-TPACK benefits from Ethics: Technically speaking, AI is the ability to learn to do a human-related activity by reasoning and interacting with the environment [59]. To put it another way, AI-powered technology can gather information from the environment and analyze it to create outputs that are environment specific and adaptable. AI is a broad term that may be analyzed using several techniques.
- H8. Intelligent-TPK benefits from Ethics: According to Zawacki-Richter et al. [79], machine learning, natural language processing, and DL are the most popular techniques. AI is integrated into a few instruments used in educational settings. Four AI-based solutions are the focus of the current study: chatbots, dashboards, automated evaluation systems, and intelligent teaching systems [39]. For instructors and students, chatbots serve as virtual or conversational agents. Educators can use text or voice input to initiate a dialogue with chatbots [51]. Educators can use chatbots to assist them in keeping students motivated [38].
- H9. Intelligent-TCK benefits from Ethics: Additionally, educators can receive updates from chatbots on their pupils' progress in learning [15]. The application of AIEd is fraught with ethical questions, notwithstanding the benefits of AI-based solutions [5, 65]. As previously indicated, AI-based technologies use their logic to do tasks. AI develops algorithms for reasoning by using historical data, sometimes known as training data.
- H10. The relationship between Intelligent ethics and TPACK is positively mediated by TCK: The AI algorithms' output will very certainly become biased since the data are biased. This means that current disparities in learning amongst student subpopulations (such as gender, race, and socioeconomic

position) might be widened by AI [34, 65]. Several ethical evaluations and considerations from various stakeholders (such as designers, developers, and end users) are necessary for responsible or explainable AI.

## 5.4 RESEARCH METHODOLOGY

### 5.4.1 Objective of the Study

Kim et al. [43] state that while educators are important members of society [51, 65], not much is known about their usage of AI-based technologies in the classroom. Research on professional knowledge evaluations for AI-based education among instructors is therefore scarce.

Furthermore, the viewpoint of the educator has not yet been examined about ethical issues and assessments of AI. As a result, nothing is known about how educators interpret and assess decisions made by AI. Finally, there is a dearth of empirical data to support the relationship between instructors' ethical judgment and their instructional AI skills. Using these shortcomings in mind, the study aims to identify educators' proficiency using AI-based technologies for TPACK-based instruction. A distinct knowledge component from PK is needed to evaluate the decisions made by AI-based technologies [34, 35]. For instance, comprehending personalized feedback chances is not the same as evaluating educators' openness and fairness. We discussed ethical evaluation as a separate area of knowledge to offer an understandable and clear viewpoint. The current study expanded on TPACK by including moral evaluations of AI-powered instruments. To put it another way, we looked for the ethical knowledge that educators ought to possess to integrate AI-based tools in an ethical manner. There could be some contributions from the current investigation. Intelligent machines also include AI-based technologies. Instructors need to be able to communicate with and operate intelligent machines [30, 31]. Thus far, a particular kind of technology has been the subject of previous studies that investigated instructors' TPACK abilities [47, 48]. In contrast to earlier research, this study focuses on the knowledge and abilities needed to engage with and use AI-based systems. An increasing amount of study is being done on how well educators embrace AI-based tools like chatbots and computerized test. We stress the incorporation of AI-based technologies into education in addition to the acceptance of AI systems. It is noteworthy that educators who lack a strong foundation in technology do not capitalize on instructional potential [54]. Their perception that technology tools are difficult to utilize is the explanation. For AI-based technologies to be used effectively, pedagogical understanding is essential. The reason for this is that AI has changed the pedagogical information that educators need to possess [63]. Some abilities for the pedagogical application of AI, such monitoring and timely intervention, were only partially understood by previous research [16, 23]. To the best of our knowledge, this is the first study to offer an integrated viewpoint on pertinent information from pedagogical and technological angles unique to AI. In some ways, instructors' use of technology integrates ethics into the mix. Except for a few research [42, 78, 68], the ethical element of TPACK has not been addressed in relation to technology use.

However, this study does not specifically address ethical issues related to AI. The new study addresses these worries by highlighting certain ethical issues with AI.

### 5.4.2 Research Design

The study utilized the descriptive methodology for understanding the TPACK model. The data is collected from academicians from education sector who use AI-enabled products or AI-powered technologies. The data collection started in January 2024 using the purposive sampling technique. The data was collected using questionnaire. The respondents belonged to central NCR region of India. The initial data collected was 247; after removing the incomplete responses, the final data resulted in 206. The data was then coded and Smart PLS software was used for applying the Structural Equation Modeling (SEM). As the data was not-normal, PLS-SEM method was utilized [32]. The scales used in the study are adopted from Celik [8], namely, TK, TCK, TPK, TPACK, and ethics. The Cronbach's alpha values for TK, TCK, TPK, TPACK, and ethics were 0.88, 0.85, 0.89, 0.91, and 0.89.

## 5.5 DATA ANALYSIS

The demographic details of the respondents are mentioned in Figure 5.2.

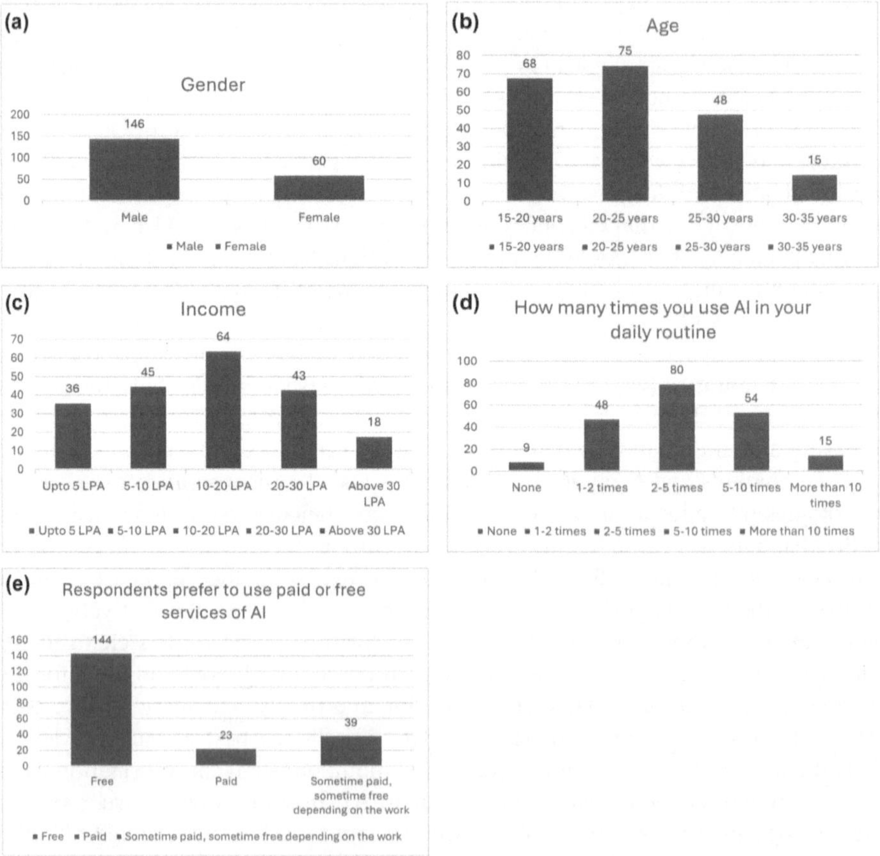


**FIGURE 5.2a–e** Demographic profile of the respondents

### 5.5.1 PLS-SEM Results

The study utilized PLS-SEM method for the purpose of testing the measurement and structural model (Figure 5.3).

The measurement model includes reliability, validity, and collinearity. Table 5.1 represents the reliability and validity. The reliability is verified as the Cronbach's alpha values of all the constructs are above more than 0.7 and composite reliability values are above 0.7 and below 0.95 [32]. The validity is verified as the average variance explained of all constructs is above 0.50 and fulfills the Fornell & Larcker criteria [32]. Thus, the data fulfills the reliability and validity criteria. Table 5.2 represents the Variance Inflation Factor (VIF) values for verifying that the data set does not possess the multi-collinearity issue. It is inferred that the data set does not have collinearity as VIF values are less than 0.5 [32].

The structural model tests the hypothesis by using the bootstrap technique. It also includes the *R* square, model fit, predictive relevance of PLS-path model, and path coefficient significance for the hypothesis formulated.

Table 5.3 represents the *R* square, model fit, and predictive relevance of the PLS-path model. The *R* square for TPACK is 0.75, which infers that the constructs explain 53% of variation in TPACK. The *Q* square represents the predictive relevance of the PLS-path model and the value for TPACK is 0.487, which infers that there is medium predictive relevance of the model. The model fit is also good as the Standardized Root Mean Square Residual (SRMR) value is less than 0.08 [32].

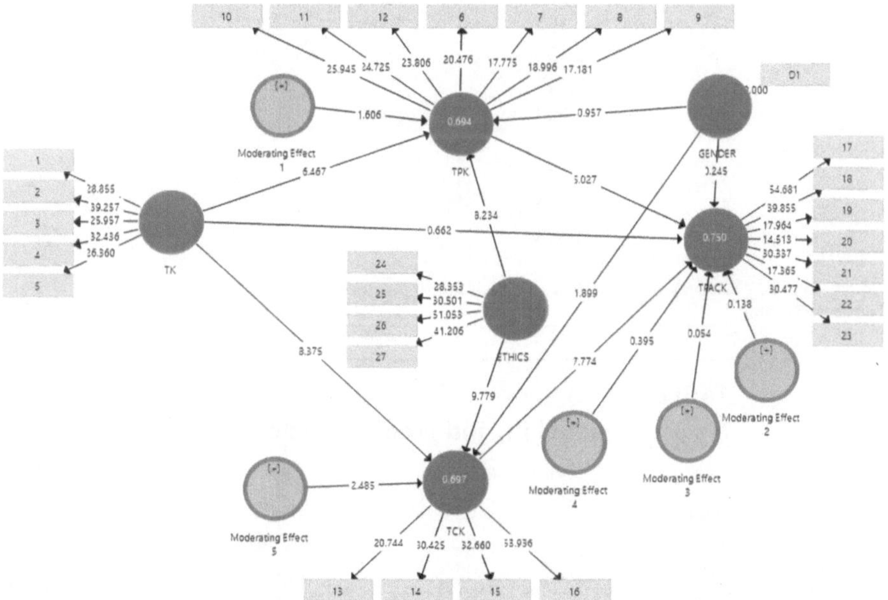


**FIGURE 5.3** PLS–Structural Equation Modeling (SEM) tested model

**TABLE 5.1**
**Reliability and validity**

| | Cronbach's alpha, CR, and AVE | | | Fornell & Larcker criteria | | | | |
|---|---|---|---|---|---|---|---|---|
| **Constructs** | **Cronbach's alpha** | **CR** | **AVE** | **Ethics** | **TCK** | **TK** | **TPACK** | **TPK** |
| Ethics | 0.89 | 0.92 | 0.75 | 0.87 | | | | |
| TCK | 0.85 | 0.90 | 0.70 | 0.76 | 0.83 | | | |
| TK | 0.89 | 0.92 | 0.69 | 0.62 | 0.72 | 0.83 | | |
| TPACK | 0.91 | 0.93 | 0.66 | 0.81 | 0.82 | 0.69 | 0.82 | |
| TPK | 0.88 | 0.91 | 0.59 | 0.76 | 0.73 | 0.73 | 0.79 | 0.77 |

*Abbreviations:* CR, composite reliability; TCK, technical content knowledge; TK, technological knowledge; TPACK, technical, pedagogical, and content knowledge; TPK, technical pedagogical knowledge.

**TABLE 5.2**
**Multi-collinearity tested**

| Items | VIF | Items | VIF |
|---|---|---|---|
| 1 | 2.324 | 15 | 1.958 |
| 2 | 3.162 | 16 | 2.512 |
| 3 | 2.385 | 17 | 3.492 |
| 4 | 2.226 | 18 | 3.266 |
| 5 | 2.023 | 19 | 2.192 |
| 6 | 2.163 | 20 | 1.765 |
| 7 | 2.106 | 21 | 2.892 |
| 8 | 2.137 | 22 | 1.999 |
| 9 | 2.047 | 23 | 2.39 |
| 10 | 2.327 | 24 | 2.216 |
| 11 | 2.207 | 25 | 2.444 |
| 12 | 2.237 | 26 | 2.733 |
| 13 | 1.688 | 27 | 2.303 |
| 14 | 1.944 | | |

*Abbreviation:* VIF, Variance Inflation Factor.

**TABLE 5.3**
***R* square, model fit, and predictive relevance**

| | ***R* square** | $Q^2$ |
|---|---|---|
| Ethics | 0.66 | 0.482 |
| TCK | 0.53 | 0.361 |
| TPACK | 0.75 | 0.487 |
| TPK | 0.54 | 0.305 |
| **SRMR** = 0.067 | | |

Table 5.4 represents the bootstrap results for testing the hypothesis. It is inferred that TK is positively linked with TPK ($\beta$=0.42, $p$<0.05) as H1 is supported at 5% level of confidence. Similarly, it is inferred that TK is positively linked with TCK ($\beta$=0.41, $p$<0.05); TK is positively linked with ethics ($\beta$=0.62, $p$<0.05); TPK is positively linked with TPACK ($\beta$=0.23, $p$<0.05); TCK is positively linked with TPACK ($\beta$=0.42, $p$<0.05); ethics is positively linked with TPACK ($\beta$=0.32, $p$<0.05); ethics is positively linked with TPK ($\beta$=0.51, $p$<0.05); ethics is positively linked with TCK ($\beta$=0.52, $p$<0.05) as H2, H4, H5, H6, H7, H8, and H9, respectively, are supported at 5% level of confidence. In contrast, TK is not positively linked with TPACK ($\beta$=0.07, $p$>0.05) as H3 is not supported. For mediation analysis, specific indirect effect is shown in the Table 5.4. It is inferred that TCK positively mediates (partially) the relationship between ethics and TPACK ($\beta$=0.19, $p$<0.05). Thus, H10 is supported partially.

Table 5.5 represents the moderating effect of gender on the relationship between ethics and TPACK; TK and TPK; and TK and TCK. It is inferred that the moderating effect of gender is only significant on the relationship between TK and TCK and not significant for the remaining relationships.

**TABLE 5.4**
**Hypothesis testing**

| Hypothesis | Beta | SD | *T* statistics (\|O/STDEV\|) | *P* values | Result |
|---|---|---|---|---|---|
| TK -> TPK | 0.42 | 0.07 | 6.26 | 0.00 | H1 supported |
| TK -> TCK | 0.41 | 0.05 | 7.69 | 0.00 | H2 supported |
| TK -> TPACK | 0.07 | 0.07 | 1.00 | 0.32 | H3 not supported |
| TK-> Ethics | 0.62 | 0.06 | 10.77 | 0.00 | H4 supported |
| TPK-> TPACK | 0.23 | 0.08 | 2.95 | 0.00 | H5 supported |
| TCK->TPACK | 0.36 | 0.07 | 5.07 | 0.00 | H6 supported |
| Ethics-> TPACK | 0.32 | 0.06 | 4.88 | 0.00 | H7 supported |
| Ethics -> TPK | 0.51 | 0.06 | 8.22 | 0.00 | H8 supported |
| Ethics -> TCK | 0.52 | 0.06 | 9.38 | 0.00 | H9 supported |
| Ethics -> TCK -> TPACK | 0.19 | 0.04 | 4.67 | 0.00 | H10 supported |

**TABLE 5.5**
**Moderating effect of gender**

| Moderating relationship | Beta | SD | *T* statistics | *P* values | Result |
|---|---|---|---|---|---|
| Ethics ->Moderating Effect 1 -> TPACK | −0.02 | 0.03 | 0.66 | 0.51 | No moderating effect |
| TK -> Moderating Effect 2 -> TPK | 0.06 | 0.04 | 1.69 | 0.09 | No moderating effect |
| TK-> Moderating Effect 3 -> TCK | 0.10 | 0.04 | 2.47 | 0.01 | Moderating effect |

## 5.6 CONCLUSION AND DISCUSSION

The study concludes that the linkage between the Intelligent-TK and Intelligent-TPK; between Intelligent-TK and Intelligent-TCK; between Intelligent-TK and ethics assessment; between Intelligent-TCK and Intelligent-TPACK; between Intelligent-TPK and Intelligent-TPACK; and between ethics assessment and Intelligent-TPACK. Thus, the use of AI has positive impact on the overall teaching process in the context of PCK and is an emerging field in the context of situational learning theory.

The study has several contributions. First, empirical testing of the Intelligent-TPACK framework for AI-enabled teaching process. Second, the ethical assessment remains elusive in the literature when it comes to TPACK framework. The study addresses this gap by including the ethical aspect in this framework. The ethical aspects improve the framework as learning is a very crucial part especially in AI-enabled education. Third, the paper adds to the body of knowledge with respect to situational learning theory for AI-enabled teaching processes. The situational learning theory is based on the concept that the pedagogy and context is important and changes as per the requirements of the situation where the teaching process is in play. The paper integrates situational learning theory for understanding the role of technology such as AI has on the teaching process.

Provided that educators possess a greater depth of knowledge to engage with AI-driven tools, they will develop a more comprehensive comprehension of the educational benefits offered by AI. Likewise, teachers are better equipped to assess the efficacy of AI since they possess a greater familiarity with AI-driven technology. Wilson et al. [75] results are like this study, and he conclude that teachers who has greater understanding of the fundamental operations of an automatic scoring system are more adept at modifying its pedagogy as per the teaching requirements. In addition, educators who have high proficiency in both fundamental and technical skills are effectively employing AI-based technologies [36]. AI helps in teaching process and makes the learning more effective [14]. Previously, TK is not found to be adequate for impact TPACK [8]. Interacting with AI technology has the capacity to improve instructors' ethical assessment skills. Enhanced proficiency in AI enables educators to critically evaluate AI-enabled products [50]. People who do not have knowledge or skills in AI may rely too much on AI systems [28]. The reason for our conclusion lies on the significant role of PK in the embracing technology. Educators achieve effective learning process by leveraging technologies that improve their expert knowledge [72]. The study emphasizes on the effective application of AI in the education sector by integrating TK with TPK, TCK, and TPACK.

Prior studies on ethical concerns with AI have primarily concentrated on the developmental facets of AI technology [19]. Nevertheless, the opinions and evaluations of end users play a vital role in the ethical implementation of AI in educational environments [34]. Teachers have a crucial role as key stakeholders in the ethical incorporation of AI [49]. Hence, the process of incorporating AI into education should encompass instructors' ethical evaluations. Nevertheless, there have been no prior studies that have specifically examined ethical evaluations based on the teacher's expertise. An important addition of the present work is the expansion of TPACK by incorporating ethical assessment information.

## 5.7 IMPLICATIONS

The study posits few practical implications for academician development. First, the training programs for academicians will emphasize on embracing technological expertise to enhance awareness for ethical assessment while using AI-enabled products. However, it is important to provide training opportunities to educators regarding pedagogical benefits of using such AI-enabled products, particularly when the goal is to improve their knowledge successful integration of AI in teaching process. Based on the findings of the study, it is recommended that educators may acquire knowledge about AI concepts and its functions for deeper understanding of the ethical issues associated with using such technologies. In this regard, technologies like Natural Language Processing (NLP), a sub-set of AI applications like chatbots, may be implemented. This will enhance the educator's knowledge regarding ethics concerns that are related to AI-driven decisions, such as transparency and accountability. In addition, the importance of pedagogical benefits offered by AI-enabled products will improve the integration process even further. Consequently, students will possess TPK, which increases the probability of using AI-enabled products in the teaching process. Thus, the field of integration of AI-enabled products in the teaching process is gaining importance and requires updating of our traditional teaching process. This will result in more enhanced learning of the students and make the teaching process more effective.

## REFERENCES

1. Akgun, Selin, and Christine Greenhow. "Artificial intelligence in education: Addressing ethical challenges in K-12 settings." *AI and Ethics* 2, no. 3 (2022): 431–440.
2. Almusharraf, Norah, and Hind Alotaibi. "An error-analysis study from an EFL writing context: Human and automated essay scoring approaches." *Technology, Knowledge and Learning* 28, no. 3 (2023): 1015–1031.
3. Brynjolfsson, Erik, and Andrew McAfee. *The Second Machine Age: Work, Progress, and Prosperity in a Time of Brilliant Technologies.* WW Norton & Company, 2014.
4. Bozkurt, Aras, Abdulkadir Karadeniz, David Baneres, Ana Elena Guerrero-Roldán, and M. Elena Rodríguez. "Artificial intelligence and reflections from educational landscape: A review of AI Studies in half a century." *Sustainability* 13, no. 2 (2021): 800.
5. Castelvecchi, Davide. "Can we open the black box of AI?" *Nature News* 538, no. 7623 (2016): 20.
6. Cavalcanti, Anderson Pinheiro, Arthur Barbosa, Ruan Carvalho, Fred Freitas, Yi-Shan Tsai, Dragan Gašević, and Rafael Ferreira Mello. "Automatic feedback in online learning environments: A systematic literature review." *Computers and Education: Artificial Intelligence* 2 (2021): 100027.
7. Celik, Ismail, Ismail Sahin, and Ahmet Oguz Akturk. "Analysis of the relations among the components of technological pedagogical and content knowledge (TPACK): A structural equation model." *Journal of Educational Computing Research* 51, no. 1 (2014): 1–22.
8. Celik, Ismail. "Towards Intelligent-TPACK: An empirical study on teachers' professional knowledge to ethically integrate artificial intelligence (AI)-based tools into education." *Computers in Human Behavior* 138 (2023): 107468.
9. Chaudhry, Muhammad Ali, and Emre Kazim. "Artificial Intelligence in Education (AIEd): A high-level academic and industry note 2021." *AI and Ethics* 2, no. 1 (2022): 157–165.

10. Chen, Xieling, Di Zou, Gary Cheng, and Haoran Xie. "Detecting latent topics and trends in educational technologies over four decades using structural topic modeling: A retrospective of all volumes of computers & education." *Computers & Education* 151 (2020): 103855.
11. Chen, Xieling, Di Zou, Haoran Xie, and Gary Cheng. "Twenty years of personalized language learning." *Educational Technology & Society* 24, no. 1 (2021): 205–222.
12. Chen, Xieling, Di Zou, Haoran Xie, Gary Cheng, and Caixia Liu. "Two decades of artificial intelligence in education." *Educational Technology & Society* 25, no. 1 (2022): 28–47.
13. Cheng, Kun-Hung, and Chin-Chung Tsai. "A case study of immersive virtual field trips in an elementary classroom: Students' learning experience and teacher-student interaction behaviors." *Computers & Education* 140 (2019): 103600.
14. Chiu, Thomas K.F., and Ching-sing Chai. "Sustainable curriculum planning for artificial intelligence education: A self-determination theory perspective." *Sustainability* 12, no. 14 (2020): 5568.
15. Chocarro, Raquel, Mónica Cortiñas, and Gustavo Marcos-Matás. "Teachers' attitudes towards chatbots in education: A technology acceptance model approach considering the effect of social language, bot proactiveness, and users' characteristics." *Educational Studies* 49, no. 2 (2023): 295–313.
16. Choi, Seongyune, Yeonju Jang, and Hyeoncheol Kim. "Influence of pedagogical beliefs and perceived trust on teachers' acceptance of educational artificial intelligence tools." *International Journal of Human–Computer Interaction* 39, no. 4 (2023): 910–922.
17. Daugherty, Paul R., and H. James Wilson. *Human+ Machine: Reimagining Work in the Age of AI.* Harvard Business Press, 2018.
18. De Cremer, David, and Leander De Schutter. "How to use algorithmic decision-making to promote inclusiveness in organizations." *AI and Ethics* 1, no. 4 (2021): 563–567.
19. Diakopoulos, Nicholas, and Michael Koliska. "Algorithmic transparency in the news media." *Digital Journalism* 5, no. 7 (2017): 809–828.
20. Dietvorst, Berkeley J., Joseph P. Simmons, and Cade Massey. "Overcoming algorithm aversion: People will use imperfect algorithms if they can (even slightly) modify them." *Management Science* 64, no. 3 (2018): 1155–1170.
21. Dogan, Selcuk, Nihan Agacli Dogan, and Ismail Celik. "Teachers' skills to integrate technology in education: Two path models explaining instructional and application software use." *Education and Information Technologies* 26 (2021): 1311–1332.
22. Dörr, Konstantin Nicholas, and Katharina Hollnbuchner. "Ethical challenges of algorithmic journalism." *Digital Journalism* 5, no. 4 (2017): 404–419.
23. Du, Yunfei, and Han Gao. "Determinants affecting teachers' adoption of AI-based applications in EFL context: An analysis of analytic hierarchy process." *Education and Information Technologies* 27, no. 7 (2022): 9357–9384.
24. Edwards, Chad, Autumn Edwards, Patric R. Spence, and Xialing Lin. "I, teacher: using artificial intelligence (AI) and social robots in communication and instruction." *Communication Education* 67, no. 4 (2018): 473–480.
25. Elliott, Anthony. *Algorithms of Anxiety: Fear in the Digital Age.* John Wiley & Sons, 2024.
26. García, Patricio, Analía Amandi, Silvia Schiaffino, and Marcelo Campo. "Evaluating Bayesian networks' precision for detecting students' learning styles." *Computers & Education* 49, no. 3 (2007): 794–808.
27. García, Patricio, Silvia Schiaffino, and Analía Amandi. "An enhanced Bayesian model to detect students' learning styles in Web-based courses." *Journal of Computer Assisted Learning* 24, no. 4 (2008): 305–315.
28. Glikson, Ella, and Anita Williams Woolley. "Human trust in artificial intelligence: Review of empirical research." *Academy of Management Annals* 14, no. 2 (2020): 627–660.

29. Guan, Chong, Jian Mou, and Zhiying Jiang. "Artificial intelligence innovation in education: A twenty-year data-driven historical analysis." *International Journal of Innovation Studies* 4, no. 4 (2020): 134–147.
30. Guggemos, Josef, and Sabine Seufert. "Teaching with and teaching about technology–Evidence for professional development of in-service teachers." *Computers in Human Behavior* 115 (2021): 106613.
31. Hair, J., W. C. Black, B. J. Babin, R. E. Anderson, and R. L. Tatham. *Multivariate Data Analysis* (6th ed.). Pearson, Upper Saddle River, NJ, 2010.
32. Hair Jr, Joe F., Matt C. Howard, and Christian Nitzl. "Assessing measurement model quality in PLS-SEM using confirmatory composite analysis." *Journal of Business Research* 109 (2020): 101–110.
33. Hinojo-Lucena, Francisco-Javier, Inmaculada Aznar-Díaz, María-Pilar Cáceres-Reche, and José-María Romero-Rodríguez. "Artificial intelligence in higher education: A bibliometric study on its impact in the scientific literature." *Education Sciences* 9, no. 1 (2019): 51.
34. Holmes, Wayne, Kaska Porayska-Pomsta, Ken Holstein, Emma Sutherland, Toby Baker, Simon Buckingham Shum, Olga C. Santos et al. "Ethics of AI in education: Towards a community-wide framework." *International Journal of Artificial Intelligence in Education,* 32 (2022): 1–23.
35. Holmes, Wayne, Maya Bialik, and Charles Fadel. *Artificial Intelligence in Education Promises and Implications for Teaching and Learning.* Center for Curriculum Redesign, 2019.
36. Holstein, Kenneth, Gena Hong, Mera Tegene, Bruce M. McLaren, and Vincent Aleven. "The classroom as a dashboard: Co-designing wearable cognitive augmentation for K-12 teachers." In *Proceedings of the 8th International Conference on Learning Analytics and Knowledge,*, 2018: 79–88.
37. Hrastinski, Stefan, Anders D. Olofsson, Charlotte Arkenback, Sara Ekström, Elin Ericsson, Göran Fransson, Jimmy Jaldemark et al. "Critical imaginaries and reflections on artificial intelligence and robots in postdigital K-12 education." *Postdigital Science and Education* 1 (2019): 427–445.
38. Huang, Weijiao, Khe Foon Hew, and Luke K. Fryer. "Chatbots for language learning—Are they really useful? A systematic review of chatbot-supported language learning." *Journal of Computer Assisted Learning* 38, no. 1 (2022): 237–257.
39. Hwang, Gwo-Jen, Haoran Xie, Benjamin W. Wah, and Dragan Gašević. "Vision, challenges, roles and research issues of artificial intelligence in education." *Computers and Education: Artificial Intelligence* 1 (2020): 100001.
40. Jang, Syh-Jong, and Meng-Fang Tsai. "Exploring the TPACK of Taiwanese elementary mathematics and science teachers with respect to use of interactive whiteboards." *Computers & Education* 59, no. 2 (2012): 327–338.
41. Jang, Syh-Jong, and Meng-Fang Tsai. "Exploring the TPACK of Taiwanese secondary school science teachers using a new contextualized TPACK model." *Australasian Journal of Educational Technology* 29, no. 4 (2013): 566–580.
42. Kabakci Yurdakul, I. Ş. I. L., and Ahmet Naci Çoklar. "Modeling preservice teachers' TPACK competencies based on ICT usage." *Journal of Computer Assisted Learning* 30, no. 4 (2014): 363–376.
43. Kim, Seonghun, Yeonju Jang, Seongyune Choi, Woojin Kim, Heeseok Jung, Soohwan Kim, and Hyeoncheol Kim. "Analyzing teacher competency with TPACK for K-12 AI education." *KI-Künstliche Intelligenz* 35, no. 2 (2021): 139–151.
44. Koehler, Matthew, and Punya Mishra. "What is technological pedagogical content knowledge (TPACK)?" *Contemporary Issues in Technology and Teacher Education* 9, no. 1 (2009): 60–70.

45. Koh, Joyce Hwee Ling, and Shanti Divaharan. "Towards a TPACK-fostering ICT instructional process for teachers: Lessons from the implementation of interactive whiteboard instruction." *Australasian Journal of Educational Technology* 29, no. 2 (2013): 233–247.
46. Koh, Joyce Hwee Ling, Ching Sing Chai, and Chin-Chung Tsai. "Examining practicing teachers' perceptions of technological pedagogical content knowledge (TPACK) pathways: A structural equation modeling approach." *Instructional Science* 41 (2013): 793–809.
47. Lave, Jean, and Etienne Wenger. *Situated Learning: Legitimate Peripheral Participation.* Cambridge University Press, 1991.
48. Lee, Min-Hsien, and Chin-Chung Tsai. "Exploring teachers' perceived self-efficacy and technological pedagogical content knowledge with respect to educational use of the world wide web." *Instructional Science* 38 (2010): 1–21.
49. Luckin, Rose, Karine George, and Mutlu Cukurova. *AI for School Teachers.* CRC Press, 2022.
50. Luckin, Rosemary, and Mutlu Cukurova. "Designing educational technologies in the age of AI: A learning sciences-driven approach." *British Journal of Educational Technology* 50, no. 6 (2019): 2824–2838.
51. Luo, Xueming, Siliang Tong, Zheng Fang, and Zhe Qu. "Frontiers: Machines vs. humans: The impact of artificial intelligence chatbot disclosure on customer purchases." *Marketing Science* 38, no. 6 (2019): 937–947.
52. Manyika, James, Michael Chui, Mehdi Miremadi, Jacques Bughin, Katy George, Paul Willmott, and Martin Dewhurst. "A future that works: AI, automation, employment, and productivity." *McKinsey Global Institute Research, Technical Reports* 60 (2017): 1–135.
53. Mishra, Punya, and Matthew J. Koehler. "Technological pedagogical content knowledge: A framework for teacher knowledge." *Teachers' College Record* 108, no. 6 (2006): 1017–1054.
54. Mishra, Punya, Matthew J. Koehler, and Danah Henriksen. "The seven trans-disciplinary habits of mind: Extending the TPACK framework towards 21st century learning." *Educational Technology* 51, no. 2 (2011): 22–28.
55. Nazaretsky, Tanya, Mutlu Cukurova, and Giora Alexandron. "An instrument for measuring teachers' trust in AI-based educational technology." In *LAK22: 12th International Learning Analytics and Knowledge Conference*, pp. 56–66, 2022.
56. Ng, Davy Tsz Kit, Jac Ka Lok Leung, Samuel Kai Wah Chu, and Maggie Shen Qiao. "Conceptualizing AI literacy: An exploratory review." *Computers and Education: Artificial Intelligence* 2 (2021): 100041.
57. Popenici, Stefan AD, and Sharon Kerr. "Exploring the impact of artificial intelligence on teaching and learning in higher education." *Research and Practice in Technology Enhanced Learning* 12, no. 1 (2017): 22.
58. Roll, Ido, and Ruth Wylie. "Evolution and revolution in artificial intelligence in education." *International Journal of Artificial Intelligence in Education* 26 (2016): 582–599.
59. Russell, Stuart J., and Peter Norvig. *Artificial Intelligence: A Modern Approach.* Pearson, 2016.
60. Sahin, Ismail, Ismail Celik, Ahmet Oguz Akturk, and Mustafa Aydin. "Analysis of relationships between technological pedagogical content knowledge and educational internet use." *Journal of Digital Learning in Teacher Education* 29, no. 4 (2013): 110–117.
61. Saini, C. P. Difficulty in transition: Study on hindrances in online shopping for potential shoppers. *International Journal of Online Marketing (IJOM)* 10, no. 2 (2020): 17–29.
62. Schmid, Mirjam, Eliana Brianza, and Dominik Petko. "Developing a short assessment instrument for Technological Pedagogical Content Knowledge (TPACK. xs) and comparing the factor structure of an integrative and a transformative model." *Computers & Education* 157 (2020): 103967.

63. Seufert, Sabine, Josef Guggemos, and Michael Sailer. "Technology-related knowledge, skills, and attitudes of pre-and in-service teachers: The current situation and emerging trends." *Computers in Human Behavior* 115 (2021): 106552.
64. Shin, Donghee, and Yong Jin Park. "Role of fairness, accountability, and transparency in algorithmic affordance." *Computers in Human Behavior* 98 (2019): 277–284.
65. Shin, Donghee. "How do users interact with algorithm recommender systems? The interaction of users, algorithms, and performance." *Computers in Human Behavior* 109 (2020): 106344.
66. Shulman, Lee S. "Those who understand: Knowledge growth in teaching." *Educational Researcher* 15, no. 2 (1986): 4–14.
67. Shulman, Lee. "Knowledge and teaching: Foundations of the new reform." *Harvard Educational Review* 57, no. 1 (1987): 1–23.
68. Smakman, Matthijs, Paul Vogt, and Elly A. Konijn. "Moral considerations on social robots in education: A multi-stakeholder perspective." *Computers & Education* 174 (2021): 104317.
69. Smutny, Pavel, and Petra Schreiberova. "Chatbots for learning: A review of educational chatbots for the Facebook Messenger." *Computers & Education* 151 (2020): 103862.
70. Tang, Jing. "Study on intelligent scoring of English composition based on machine learning from the perspective of natural language processing." *ACM Transactions on Asian and Low-Resource Language Information Processing,* December 2020 Proceedings (2020): 289–294.
71. Thomaz, Andrea L., and Cynthia Breazeal. "Teachable robots: Understanding human teaching behavior to build more effective robot learners." *Artificial Intelligence* 172, no. 6–7 (2008): 716–737.
72. Tondeur, Jo, Dominik Petko, Rhonda Christensen, Kerstin Drossel, Louise Starkey, Gerald Knezek, and Denise A. Schmidt-Crawford. "Quality criteria for conceptual technology integration models in education: Bridging research and practice." *Educational Technology Research and Development* 69 (2021): 2187–2208.
73. Valtonen, Teemu, Erkko Sointu, Jari Kukkonen, Sini Kontkanen, Matthew C. Lambert, and Kati Mäkitalo-Siegl. "TPACK updated to measure pre-service teachers' twenty-first-century skills." *Australasian Journal of Educational Technology* 33, no. 3 (2017): 15–31.
74. Wang, Youmei, Chenchen Liu, and Yun-Fang Tu. "Factors affecting the adoption of AI-based applications in higher education." *Educational Technology & Society* 24, no. 3 (2021): 116–129.
75. Wilson, Joshua, Cristina Ahrendt, Emily A. Fudge, Alexandria Raiche, Gaysha Beard, and Charles MacArthur. "Elementary teachers' perceptions of automated feedback and automated scoring: Transforming the teaching and learning of writing using automated writing evaluation." *Computers & Education* 168 (2021): 104208.
76. Xu, Ling. "The dilemma and countermeasures of AI in educational application." In *Proceedings of the 2020 4th International Conference on Computer Science and Artificial Intelligence*, pp. 289–294, 2020.
77. Yuan, Kuo-Ching, Lung-Wen Tsai, Ko-Han Lee, Yi-Wei Cheng, Shou-Chieh Hsu, Yu-Sheng Lo, and Ray-Jade Chen. "The development an artificial intelligence algorithm for early sepsis diagnosis in the intensive care unit." *International Journal of Medical Informatics* 141 (2020): 104176.
78. Yurdakul, Isil Kabakci, Hatice Ferhan Odabasi, Kerem Kilicer, Ahmet Naci Coklar, Gurkay Birinci, and Adile Askim Kurt. "The development, validity and reliability of TPACK-deep: A technological pedagogical content knowledge scale." *Computers & Education* 58, no. 3 (2012): 964–977.
79. Zawacki-Richter, Olaf, Victoria I. Marín, Melissa Bond, and Franziska Gouverneur. "Systematic review of research on artificial intelligence applications in higher education–where are the educators?" *International Journal of Educational Technology in Higher Education* 16, no. 1 (2019): 1–27.

# 6 Marketing Automation Development as a Process of Digital Ecosystem Advancement

*Farshad Badie, Palanivel Rathinasabapathi Velmurugan, and Ajay Kumar*

## 6.1 BACKGROUND OF THOUGHT

My point of departure is a special focus on the concept of *production*. Production is an operational process that involves the combination of different inputs, such as human beings, machines, materials, and monetary resources. All these inputs are used to create specific outputs (i.e., *products*) that can be either *goods* (i.e., tangible, separable, and homogenous products) or *services* (i.e., intangible, inseparable, and heterogeneous products). Any production process can be interpreted as a complex and dynamic 'system' that requires careful management to ensure that the right inputs are combined in the most effective way to create the desired outputs. For more information on products and production processes, see Womack et al. (1990), Kaplan & Norton (1996), Schmenner (2004), Stevenson (2009), and Slack et al. (2016).

Relying on the process of 'production' and the final 'products' (of a production system), we can fundamentally define the concept of *marketing*. In general words, marketing is a systematic process aimed at identifying, characterizing, promoting, and advertising a particular product to potential customers (i.e., *users*) in some specific environment [McCarthy (1960), Kotler & Keller (2022), Verhoef et al. (2003), Solomon et al. (2023)].

In the digital age, we have had significant developments in *marketing automation* [Hoffman et al. (2022)] and, in fact, in making marketing processes automatic based on information-based and software-based technologies. This chapter will conceptually deal with the concept of 'marketing' and will attempt to, systematically and analytically, become concerned with the concept of 'marketing automation.' Equivalently, the main contribution of this chapter is to systematically analyze 'marketing automation' and structurally existentially analyze how, and why, we can interpret 'marketing automation development' as a dynamic process of *digital ecosystem* advancement. In the framework of this research, as it will be analyzed in this chapter, a digital ecosystem [Osorio (2023)] is a network of interconnected digitalized

 DOI: 10.1201/9781032700953-7

technologies and frameworks that organizationally work together to create value for users and businesses.

## 6.2 WHAT IS MARKETING?

I shall start this section, and my structural analysis of the concept of 'marketing,' with a special systematic focus on five important building blocks of marketing (so-called *Marketing Mix*). The marketing mix includes *product*, *price*, *place*, *promotion*, and *people* [McCarthy (1960), Dibb et al. (2009), Lamb et al. (2020)].

*Product* is the first element of the marketing mix and, as characterized above, it refers to the goods or services that an organization (i.e., producer) offers to its customers/users. In my opinion, product is the core (i.e., the most important element) of the marketing mix. The reason is that all other elements depend on it. It is important for a company to understand its product's existence, features, benefits, and unique selling proposition (USP) in order to be able to introduce it to the world as well as to develop an effective marketing strategy for it. A scientific analysis of any specific product involves evaluating its design, features/characteristics, quality, and functionality. It is also essential to evaluate the target market's needs and preferences to ensure that the product meets their requirements. Companies must continuously innovate and, correspondingly, improve their products to stay competitive in the market.

*Price* is the second building block of the marketing mix. It refers to the amount of money a customer pays to purchase or, to be involved in the usage of, some specific product. Price can be regarded as a critical element because it affects the company's profitability and market share. Correspondingly, *pricing* is a significant function that is created based on the critical element 'price.' Pricing strategies can vary depending on the product, competition, and target market. From a scientific point of view, the static concept 'price' and the dynamic concept 'pricing' are concerned with the evaluation of the product's perceived value, production costs, and competitive landscape. Companies must set a price that reflects the product's quality and value while remaining competitive in the market. Pricing strategies such as discounts, bundling, and dynamic pricing are always being utilized to maximize companies' profits.

*Place* refers to the location where customers can purchase and be served the product. Place involves the distribution channels used to deliver the product to the customer, such as retail stores, e-commerce websites, or direct sales. Place is an essential component of the marketing mix, as it affects the availability and accessibility of the product. Scientifically speaking, I can state that marketing, by dealing with 'place,' would be dealt with the evaluation of the target market's purchasing habits and preferences. Companies need to have a special focus on distribution channels that are convenient as well as accessible to their target market. They also attempt to ensure that their products are available at the right place and at the right time to maximize sales.

*Promotion* describes the communication strategies that are organized and plotted to promote the product to the target market. Promotion involves advertising, sales promotion, public relations, and personal selling. It is also a critical element of the marketing mix. The main reason is that it supports the creation of awareness based on, as well as of interest in, some specific product. Focusing on 'promotion' scientifically, we can understand that any dynamic concept of 'promotion' involves the evaluation of the target market's communication preferences and media habits. Companies must choose right communication channels that are effective in reaching their target market. They must also create messages that are compelling and resonate with their audience.

*People* mainly refer to the employees who represent the company and interact with customers. People are very important because they can influence the customer's perception of the product as well as of the company. Scientifically assessing, companies must ensure that their employees have the necessary knowledge, skills, and abilities to provide excellent customer service. They must also create a culture that can promote customer-centricity and also foster a positive attitude towards the customer.

Marketing involves several key stages, including product research and development, branding, pricing, promotion, and distribution. In each of these stages, marketers employ various tools and strategies to effectively communicate the product's USP [Ries & Trout (2001), Trout & Rivkin (2009)] and value proposition to the target audience. To succeed in marketing, it is essential to have a deep understanding of the target market, including their preferences, behaviors, and cultural and individualistic values. By understanding these factors, marketers can create marketing *campaigns* [Schultz et al. (2014), Duncan & Moriarty (2017)] that resonate with the target audience and persuade them to take action.

At this point, it shall be stressed that marketing has (and, in fact, all fundamental elements of marketing have) a special emphasis on understanding humans' *needs* (based on their individualistic, as well as on their societal/cultural, *values*). In fact, we can interpret that 'values' are the most influential concepts that can affect consumers' purchasing and using decisions.

## 6.3 WHAT DOES MARKETING AUTOMATION MEAN?

In the digital age, we have had significant developments in the automation of marketing (or *marketing automation*) processes and, in fact, in utilizing technologies (mainly based on information, computing, and software sciences) toward the management of different marketing scenarios and processes. In general words, marketing automation refers to how software and technology can be used to streamline and automate various marketing tasks and processes. Marketing automation allows businesses to improve their efficiency, save time, and enhance the customer experience [Gupta & Lehmann (2003), Verhoef (2013), Eriksson et al. (2015), Pantano & Bachman (2016)].

Marketing automation can be utilized to manage various marketing activities (e.g., email marketing, social media marketing, lead generation, and customer segmentation). For example, in email marketing, businesses can use marketing automation tools to send targeted and personalized emails to subscribers based on their individualized needs, preferences, behaviors, and purchase history. These tools do also automate the process of lead generation by collecting data and tracking user behavior to identify potential customers. Marketing automation is also being used to improve the customer experience. For instance, businesses can use chatbots or virtual assistants to interact with customers, answer their queries, provide personalized recommendations, and provide personalized and customized feedback. Marketing automation technology does also support businesses in providing a more engaging and interactive experience to customers/users while freeing up human resources for more complex tasks. Moreover, marketing automation measures and analyses the effectiveness of marketing campaigns. By collecting and analyzing data, businesses would be able to identify the most effective marketing strategies and, subsequently, to make informed decisions about future marketing efforts.

I shall come up with the statement that marketing automation—in order to allow businesses to efficiently manage and optimize their marketing strategies—simplifies and automates repetitive marketing tasks/functions. More technically, marketing automation utilizes data-driven insights and analytics to personalize and target marketing campaigns, resulting in improved customer engagement and higher conversion rates. This technology enables businesses to deliver relevant and timely content to their audience across various channels. Moreover, it aids in lead nurturing and scoring, assisting businesses in identifying and prioritizing high-quality leads for their sales teams. Such an effective functionality of marketing automation can ultimately increase revenues as well as business growth. So, undoubtedly, with its ability to automate marketing tasks, optimize marketing efforts, and enhance customer interactions, marketing automation has become an indispensable framework for various businesses to stay competitive and achieve marketing success in our digitalized world.

## 6.4 MARKETING AUTOMATION AND DEALING WITH HUMAN VALUES

As pointed out above, any marketing function at core needs to understand humans' individualistic, cultural, and societal values. So, this important essence must be taken into account by 'marketing automation' systems. In this section, I will be more specific on human values as well as on the reflections of values in automation systems.

### 6.4.1 What Are Human Values?

In this section, I am going to focus on the heuristic question of 'what values are?' I shall first review some of the most significant research works in 'value' analysis and then rely on my definition.

Peters & Waterman (1982) have identified eight key management principles that contribute to great companies' success. Among these principles are a focus on action, a close connection to customers, and respect for people. These principles emphasize the importance of aligning company values with employee empowerment and customer focus, ultimately leading to superior performance. Penrose (1959/1995) explores the internal dynamics of a firm's growth. Penrose argues that firms grow by utilizing their existing resources and capabilities to pursue new opportunities. According to Penrose, a key concept is the importance of a firm's 'value constellation,' which encompasses its knowledge, skills, and assets. Thereby, aligning these internal values with market demands is crucial for sustainable growth. Porter & Kramer (2011) propose the concept of 'Creating Shared Value,' arguing that businesses can generate both economic profits and social benefits. Porter and Kramer suggest that companies need to identify and address societal challenges that intersect with their core operations. So, they will be able to create value for both shareholders and society as a whole. Porter and Kramer emphasize aligning business values with social good to achieve a win–win situation.

In this research, I am going to rely on my definition of 'values.' As I have analyzed and argued in Badie (2024), values can be regarded as some qualitative concepts with, and based on, which human beings construct *meanings* for themselves in different contexts [read more about my definition and analysis of the dynamic concept of 'meaning' in Badie (2017a–b, 2018, 2020a–b, 2021)]. Consequently, it can be interpreted that we can—by assessing human beings' values—evaluate (and, in some cases, measure) human behaviors, activities, goals, and situations in different contexts of the world.

There is no doubt that modern technologies have had a significant impact on the way people communicate, work, and live their daily lives. With the increasing adoption of technologies such as social media, messaging applications, and online platforms, people have become more connected and are able to share and access information more easily than ever before. These technologies have also enabled individuals to express themselves in new and creative ways and to participate in a wide range of social and cultural activities. More specifically, in the context of information-, and software-, based technologies, human beings are expressing themselves, getting connected with each other, and carrying out their daily activities using technologies. So, it shall be noted that modern technologies need to acknowledge and address the values that are created and developed by human beings.

However, it is important to recognize that technologies, themselves, do not have values. Rather, they are humans who create and utilize technologies with their values. Consequently, it is vital for designers and developers of contemporary technologies to understand the potential psychological, societal, and cultural repercussions of their inventions and to take measures to guarantee that their innovations are produced and employed in an ethical, logical, cognitive, and accountable manner. This involves recognizing the potential biases and limitations of their creations and actively working to overcome them to prevent harm to individuals and societies. By doing so, designers and developers can ensure that their technologies serve to promote the well-being of humanity rather than perpetuate inequality and injustice.

### 6.4.2 Human Values in Marketing Automation Processes

As 'marketing automation' enables marketers to better understand and engage with their audience in a more personalized and targeted way, it can play a significant and critical role in acknowledging and addressing the values that are created by humans. By utilizing technology to automate and streamline various marketing processes, companies collect and analyze data about their customers'/users' preferences and behaviors (and, in fact, about their values). This information will be used to tailor marketing messages and campaigns to better align with the needs and interests of their target audience.

However, it is important to note that marketing automation does not independently support all the challenges and risks associated with modern technologies. While it can help companies to connect with their audience in a more meaningful way, it is also important to ensure that these technologies are used responsibly and ethically. This includes respecting customers'/users' privacy and data rights, and ensuring that marketing messages are truthful, are not misleading, and are transparent about the systematic use of data as well as about how data is being (or is going to be) strategically used. Furthermore, marketing automation should be viewed as a framework to support (and, not, be viewed as a replacement for) human creativity and decision-making. While technology can provide valuable insights and automate certain processes, it can never replace the human touch and the importance of building genuine relationships with customers/users.

### 6.4.3 Marketing Automation as a Process of Digital Ecosystem Advancement

A digital ecosystem refers to a group of interconnected and integrated information-based and computational components that work together to achieve a common objective or mission [for more information about 'digital ecosystems,' see Moore (1996), Iansiti & Levien (2004), Thomas et al. (2014), and Gassmann et al. (2017)].

In my opinion, the creation and development of any 'marketing automation' is a specific sub-process of digital ecosystem advancement. To properly analyze marketing automation development as a digital ecosystem advancement, it is important to focus on the 'human' element as well as on how human values come into play in information-/software-based scenarios and systems. In other words, we need to take into account the significant role that humans play in the development and implementation of marketing automation strategies, and also how these strategies are influenced by the values and beliefs of individuals and society as a whole. So, I shall summarize that a digital ecosystem is a collection of interrelated and integrated information-based components that work together to achieve a common goal or mission (that has been conceptualized and defined based on human values). In addition, the process of the development of a marketing automation system is a sub-process of digital ecosystem advancement.

As I will be more specific in the detailed analysis, marketing automation can be interpreted to be structurally developed in six interconnected layers (i.e., *Market Research*, *Production Analysis*, *Contextualization*, *Digitization*, *Computerization*,

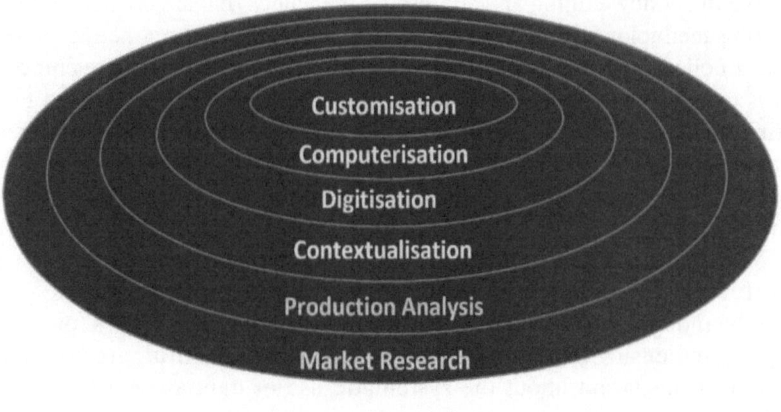


(Self-Created)

**FIGURE 6.1** A conceptual model of marketing automation development

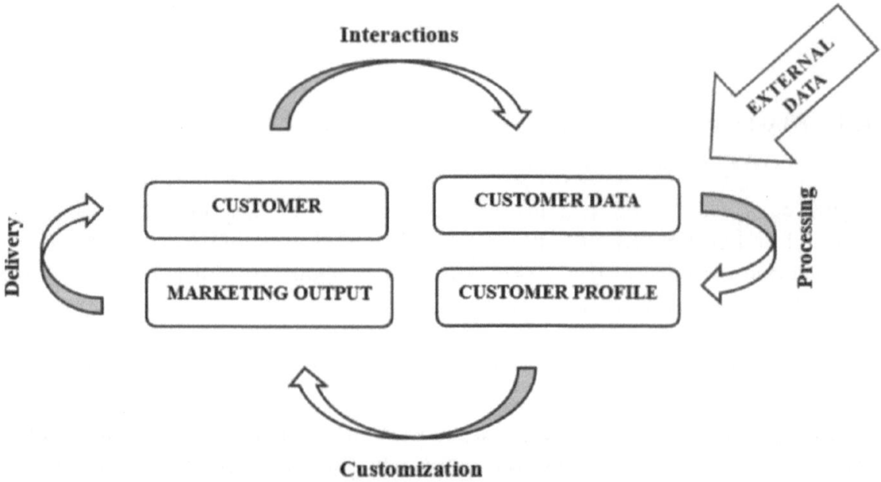


(Self-Created)

**FIGURE 6.2** A process of personalization

and *Customization*). Figure 6.1 represents the analyzed layers of marketing automation development and their interrelationships. The following sections focus on the detailed analysis of different layers of marketing automation (Figure 6.2).

## 6.5 MARKET RESEARCH (FIRST LAYER)

*Market research* fundamentally deals with the technological and digital 'landscape' (as well as 'environment') that encompasses various components such as hardware devices, software applications, communication networks, and data storage systems. Market research is also concerned with the Internet, cloud computing infrastructure, and emerging technologies [such as Artificial Intelligence (AI), the Internet of Things (IoT), and blockchain].

At this point, I need to focus on the fundamental question of 'what market research believes in and focuses on?' It can be realized that market research believes that the environment of our digital ecosystem is constantly evolving and that new technologies and digital tools/frameworks are emerging at an unprecedented rate. Such a belief is in line with how highly dynamic and complex digital environments require different businesses and organizations to be 'agile' (and, in fact, 'adaptable') to stay competitive [Gioia et al. (2000), Eisenhardt et al. (2010), Nadkarni and Wadhwa (2007), Verhoef (2013), Teece (2018)].

It is extremely important to be taken into account that market research has a special interconnectedness with human 'participation' (and 'involvement') in the technological world. In my view, participation and involvement are both the consequences of human values (that are created by individual customers/users and societies). So, it is significantly important to recognize the interconnection between 'market research' and 'human participation' (especially, based on human values) in the modern information-based technological world. This means that market research cannot exist in isolation from human behavior and the evolving technological landscape.

Market research involves gathering and analyzing data that are related to consumer/user behavior, preferences, and needs. These data are used by businesses to make informed decisions about their products and marketing strategies. However, in today's world, where technology is ubiquitous, it is crucial to take into account how people interact with technology and how technology affects their behavior. For instance, consumer behavior is heavily influenced by social media platforms, search engines, and e-commerce websites. These platforms collect vast amounts of data about customers/users. These data are collected to be leveraged by businesses for market research purposes. Similarly, all other relevant emerging technologies (especially, AI, IoT, and blockchains) are changing the way people interact with different kinds of products. So, the styles and the hownesses of the interactions of (and interrelationships between) customers/users and the products do also need to be taken into account in market research.

It is noticeable that the interconnection between market research and human participation in the technological world extends beyond data collection. It also involves considering ethical considerations (especially, regarding data privacy, data security, and data transparency). As businesses collect and use more data, there will be a more need to ensure that customers'/users' rights and interests are protected.

Here I should summarize that the first layer has a special concentration on the interrelationships between minds and environments and, subsequently, on how the concepts of 'participation' and 'involvement' would be seen, interpreted, and analyzed in our ecosystem.

## 6.6 PRODUCTION ANALYSIS (SECOND LAYER)

The second layer expresses *production analysis*. Production analysis is concerned with how the concept of 'production' can be seen, characterized, identified, described, and analyzed. Productions are some dynamic functions (i.e., transformations) from 'resources' into 'products.' Thereby, relying on our 'market research' (i.e., layer 1), we can analyze how some resources could—existentially, structurally, and strategically—be transformed into some specific product. Production analysis models are used by businesses to evaluate the efficiency and effectiveness of their production processes [Sihn (2006), Taj et al. (2011), Slack et al. (2016), Melnyk & Gottschalk (2020]. In some more specific words, by analyzing the various stages of a production process, businesses can identify areas for improvement and optimize their resource utilization.

It shall also be noted that market research (i.e., layer 1) provides valuable insights into how resources can be transformed into products. By understanding what customers/users want (i.e., layer 1), businesses can determine how resources should be allocated to meet those needs. This involves considering the *existential*, *structural*, and *strategic* factors that impact the production process. Considering some specific product (say P), 'existential factors' refer to the fundamental nature of P and P's purpose. For instance, a business that produces eco-friendly products may need to allocate more resources towards sustainability and ethical sourcing. 'Structural factors' refer to the physical and organizational structures that support the production process. For example, a business may need to invest in a new technology or equipment (say a modern LED) to improve the efficiency of its production processes. 'Strategic factors' refer to the broader goals and objectives of the business. For example, a business may need to consider factors such as market competition, pricing, and marketing strategies when allocating resources toward product development. By analyzing the aforementioned factors, businesses can determine how best to allocate their resources to optimize their production processes. This involves transforming raw materials, labor, and all other resources into finished products that meet the needs and expectations of consumers. And we should not forget that expectations have roots in 'values.'

## 6.7 CONTEXTUALIZATION (THIRD LAYER)

The third layer, or *contextualization*, deals with the identification and analysis of digital platforms. Note that there is a strong correlation between *contextualization* (of any production process) and *industrialization* (based on that specific production process). In more specific words, our contextualization and, correspondingly, our industrialization, would be supported by how we can observe 'production analysis' (i.e., layer 2) and 'market research' (i.e., layer 1) from the perspectives of each other.

'Contextualization' refers to the process of conceptualizing, interpreting, and understanding a phenomenon within some specific domain. Relying on the dynamic concepts of production analysis (i.e., layer 2) and market research (i.e., layer 1), contextualization involves analyzing the various factors that influence the production process and, also, the market in which products are offered/served. This includes understanding consumer behavior, preferences, and needs, as well as the technological and organizational structures that support the production process.

'Industrialization,' on the other hand, refers to the process of transforming raw materials and resources into finished products using technology and machinery.

Industrialization involves optimizing the production process to increase efficiency and reduce costs, while also ensuring that the final products meet the needs and expectations of customers/users.

As mentioned above, contextualization and, correspondingly, industrialization can be supported by observing production analysis (i.e., layer 2) and market research (i.e., layer 1) from the perspectives of each other. In other words, when analyzing the production process, by considering the insights gained from market research, businesses can optimize their production processes to better meet humans' needs. Similarly, by considering the insights gained from production analysis when conducting market research, businesses can gain a better understanding of the factors that impact consumer behavior and preferences.

## 6.8 DIGITIZATION (FOURTH LAYER)

*Digitization* attempts to transform industrial, and contextual, values (as the outcomes of our contextualization of production) into digits and, correspondingly, into digital components.

Digitization is responsible for the digital representation and modeling of our contexts (that are provided and developed in layer 3). In general words, digitization refers to the process of transforming analog information, such as paper documents or physical products, into 'digital' formats that can be easily stored, shared, and analyzed using digital technologies. More particularly, in our domain of interest, digitization enables companies to streamline their operations and improve their efficiency by automating processes. This will reduce manual labor and increase the speed and accuracy of data analysis.

The outcomes of our contextualization (i.e., layer 3) of production (i.e., layer 2) refer to the insights and knowledge that companies can gain by analyzing the digital components of their production processes. By analyzing data from various sources (i.e., layer 1), such as customer feedback, sales data, and production metrics, companies can gain valuable insights into their operations, identify areas for improvement, and make data-driven decisions to optimize their performance.

Digitization is a key driver of transformation in the marketing industry. It enables companies to capture and analyze contextual values and gain valuable insights into their operations and operational processes. By incorporating digital technologies into their operations, companies can improve their efficiency, competitiveness, and overall performance.

## 6.9 COMPUTERIZATION (FIFTH LAYER)

Relying on 'digitization,' the fifth layer, or *computerization* constructs a platform for computing and analyzing digital components. In general, 'computerization' refers to the process of automating manual tasks or processes using computers or other digital technologies, while 'digitization' refers to the conversion of analog information into digital formats that can be processed by computers [Lacity & Willcocks (2017)].

In digital marketing domains, computerization refers to the use of digital technologies to automate marketing processes and tasks (mainly, customer segmentation, email marketing, social media management, and digital advertising). Digitization, on the other hand, refers to the conversion of analog marketing data into digital formats that can be processed by computers (here, customer information, sales data, and website analytics are very good examples). It shall be emphasized that digitization (i.e., layer 4) supports computerization (i.e., layer 5) in the framework of 'digital marketing development' in several ways.

Digitization allows for the collection and analysis of digital marketing data. It can inform and automate marketing decisions. For example, in this framework, data on customer behavior, marketing performance, and campaign optimization are trackable and analyzable. Subsequently, marketers will be enabled to make data-driven decisions, automate different marketing campaigns, and optimize strategies for improved outcomes.

In addition, digitization enables the automation of significant marketing processes, e.g., email marketing and social media management. By digitizing customer data, it becomes possible to segment customers based on their preferences and behaviors (and, based on their internal values) as well as to support the development of targeted and personalized marketing campaigns. Moreover, in such a framework, digitization supports the automation of marketing activities (like posting on social media and digital advertising) and it can reduce manual labor and improve the efficiency and effectiveness of marketing operations.

Moreover, digitization supports the integration of different digital marketing systems and technologies, and the development of integrated marketing automation platforms. By digitizing data and processes, we will be able to integrate different marketing systems (e.g., customer relationship management software, email marketing platforms, and social media management tools) into a single, unified, platform.

## 6.10 CUSTOMIZATION (SIXTH LAYER)

*Customization*, which is the sixth and final layer, deals with how our analyzed digital components (based on layer 5, 'computerization') can be related to specified human needs, demands, and values. Customization refers to the process of tailoring products or services to meet the specific needs and demands of individual customers/users. In the domain of digital marketing systems, customization involves using data and analytics to personalize marketing messages, offers, and experiences to individual customers based on their preferences and behavior.

The analysis of digital components, such as website analytics, social media metrics, and customer/user data, can help digital marketers identify patterns and trends in customer/user behavior and preferences. This information can then be used to create personalized marketing messages, and offers, that are tailored to individual customers/users. For example, a digital marketer may use website analytics to identify the pages and products that a customer/user has viewed on their website. They may then use this information to send targeted email campaigns or retargeted digital advertisements to the customer/user, promoting products or offers that are related to their previous browsing behavior. This approach increases the likelihood that the customer/user will engage with the marketing message and take action (for example, purchase).

**TABLE 6.1**
**Difference between traditional and digital inbound marketing**

| Particulars | Traditional marketing | Digital inbound marketing |
|---|---|---|
| Basis | Interruption | Organic |
| Focus | Finding customers | Getting found by potential, existing, and aspirational customers |
| Aim | Increased sales | Creating long-lasting relationships by reaching and converting qualified customers |
| Target | Large audiences | Interested prospects |
| Tactics | Printed advertisements<br>TV advertisements<br>Outdoor advertising<br>Cold calling<br>Trade shows<br>Email lists | Blogs, e-books, whitepapers, videos on YouTube, etc.<br>SEO tactics, infographics, webinars, feeds, rich site summary, and social media marketing tactics. |

Moreover, customization can be used to create personalized customer experiences that meet the unique needs and preferences of individual customers/users. For example, a digital marketer may use data and analytics to personalize the content and layout of a website or mobile application based on the customer's/user's preferences and behavior. Increasing the likelihood that the customer/user will remain loyal to the brand and repeat purchases, this approach creates a more engaging and relevant customer/user experience.

Customization is the deepest function of marketing automation based on which the concept of 'marketing' becomes manifested in computer-based tools and software (Table 6.1).

## 6.11 CONCLUDING REMARKS

In this chapter, I have conceptually addressed the concept of 'marketing' and have respectively dealt with the analytical exploration of 'marketing automation.' Marketing automation, by empowering marketers to gain deeper insights and connect with their audience on a personalized and focused level, can have a substantial and pivotal impact on recognizing and responding to human-created values. Using technology to automate and optimize different marketing tasks, companies can gather and analyze data about their customer's preferences, behaviors, and values.

The primary contribution of this chapter has been the systematic analysis of marketing automation, specifically in terms of viewing it as a dynamic process of digital ecosystem advancement. According to this research, marketing automation development is a process of digital ecosystem advancement and is structurally equivalent to 'customization of the computerization of the digitized contextualized productions in various environments (and based on different needs and values).'

Moreover, in this chapter, I have had a special focus on 'digital transformation.' I have acknowledged the fact that marketing automation should be used to recognize and respond to human-created values. I have also analyzed marketing automation's impact on business strategies in the digital age. In more specific words, the chapter has:

1. Explored the intersection of marketing, technology, and societal values within a digital ecosystem. This interdisciplinary approach directly contributes to the valuable focus of digital technologies on how digitalization impacts business and society.
2. Critically analyzed marketing automation, not just as a marketing framework, but as a force shaping the digital ecosystem. This aligns with the theoretical and critical analysis of digitalization.
3. Dealt with marketing automation's role in digital ecosystem advancement, that is an evolving concept in the modern world.

## REFERENCES

Badie, F. 2017a. "On Logical Characterisation of Human Concept Learning Based on Terminological Systems." *Logic and Logical Philosophy* 27, no. 4: 545–566.

Badie, F. 2017b. *A Theoretical Model for Meaning Construction through Constructivist Concept Learning: A Conceptual, Terminological, Logical and Semantic Study within Human-Human-Machine Interactions*. Aalborg Universitetsforlag. https://doi.org/10.5278/vbn.phd.hum.00073.

Badie, F. 2018. "A Description Logic Based Knowledge Representation Model for Concept Understanding." In *Agents and Artificial Intelligence. ICAART 2017 (Selected Revised Papers)*, Lecture Notes in Computer Science, vol. 10839, edited by J. van den Herik, A. Rocha, and J. Filipe, 1–21. Cham: Springer.

Badie, F. 2020a. "A Formal Ontology for Conception Representation in Terminological Systems." In *Reasoning: Logic, Cognition, and Games*, Studies in Logic and Practical Reasoning, edited by M. Urbanski, T. Skura, and P. Lupkowski, 137–157. London: College Publications.

Badie, F. 2020b. "Logic and Constructivism: A Model of Terminological Knowledge." *Journal of Knowledge Structures & Systems* 1, no. 1: 23–39.

Badie, F. 2021. "Nominal Conceptualism and Logical Modelling of Agents' Conceptions." *Логико-философские штудии* 19, no. 1: 95–100.

Badie, F. 2024. *From Iceberg to Insights: A Human-Centric Guide to Competency in the AI Age*. Magestic Technology Solutions (P) Ltd. ISBN: 978–93–92090-96–7 (Print), 978–93–92090-51–6 (Online). DOI: www. doi.org/10.47716/978-93-92090-51-6.

Dibb, S., Simkin, L., Pride, W. M., and Ferrell, O. C. 2009. *Marketing Concepts and Strategies*. 6th ed. Houghton Mifflin Harcourt Publishing Company, United States.

Duncan, T., and Moriarty, S. E. 2017. *Creating Effective Advertising*. Sage Publications, United States.

Eisenhardt, K. M., Graffin, S. E., Burgelman, R. A., and Furr, N. 2010. "Strategy as Simple Rules, Fraught with Complexity in Entrepreneurial Firms." *Strategic Management Journal* 31, no. 11: 1116–1139.

Eriksson, D., Johansson, M., and Mosak, L. 2015. *Marketing Automation: A Strategic Approach to Streamline Marketing Processes*. KTH Royal Institute of Technology, Sweden.

Gassmann, O., Frankenberger, K., and Cseri, G. 2017. *Business Model Innovation: An Introduction*. Cheltenham: Edward Elgar Publishing.

Gioia, P., Schultz, M., and Corley, K. M. 2000. "Organizational Identity, Configuration, and Adaptation: Overcoming Inertia and Responding to Environmental Change." *Strategic Management Journal* 21, no. 8: 835–863.

Gupta, P., and Lehmann, D. R. 2003. "Customer Experience Management in Marketing." *Journal of Marketing Research* 40, no. 3: 390–400.

Hoffman, D. L., Novak, T. P., and Yung, Y. 2022. "Serving Hyperconnected Customers through Marketing Automation." *Journal of Interactive Marketing* 62: 137–155.

Iansiti, M., and Levien, R. 2004. *Keystone Advantage: How to Unlock the Hidden Power of Ecosystem Economics*. Harvard Business School Press, United States.

Kaplan, R. S., and Norton, D. P. 1996. *The Balanced Scorecard: Measures for Driving Performance*. Harvard Business School Press, United States.

Kotler, P., and Keller, K. L. 2022. *Marketing Management*. 16th ed. Pearson Education, India.

Lacity, M. C., and Willcocks, L. P. 2017. *Digital Transformation: A Primer for Managers*. Routledge, United Kingdom.

Lamb, C. W., Hair, J. F., McDaniel, C., and Hair Jr., J. F. 2020. *Marketing*. 11th ed. Cengage Learning, United States.

McCarthy, E. J. 1960. *Basic Marketing: A Managerial Approach*. Richard D. Irwin, Homewood, IL.

Melnyk, B. V., and Gottschalk, P. 2020. *Operations Management: A Logistics Perspective*. 12th ed. McGraw-Hill Education, New York.

Moore, J. F. 1996. *The Death of Competition: Leaders and Followers in the New Digital Economy*. Harper Business, New York.

Nadkarni, S., and Wadhwa, R. 2007. "The Strategic Management of Technological Innovation: Balancing Technological Leadership, Market Knowledge, and Organizational Capabilities." *Strategic Management Journal* 28, no. 7: 777–800.

Osorio, T. L. 2023. *Digital Business Ecosystems*. Lorenzo Tural Osorio, Germany.

Pantano, E., and Bachman, R. 2016. *The Marketing Automation Handbook: Learn How to Use Marketing Automation to Achieve Your Marketing Goals and Reduce Marketing Costs*. John Wiley & Sons, United States.

Penrose, E. T. (1959/1995). *The Theory of the Growth of the Firm*. Oxford: Oxford University Press.

Peters, T. J., and Waterman, R. H. 1982. *In Search of Excellence: Lessons from America's Best-Run Companies*. New York: Harper & Row.

Porter, M. E., and Kramer, M. R. 2011. "Creating Shared Value." *Harvard Business Review* 89, no. 1/2: 60–77.

Ries, A., and Trout, J. 2001. *Positioning: The Battle for Your Mind in a Saturated Marketplace*. McGraw-Hill, New York.

Schmenner, E. W. 2004. "Manufacturing Based Strategy for Global Competition." *Business Horizons* 47, no. 6: 19–31.

Schultz, D. E., Tannenbaum, B., and Lauterborn, R. 2014. *Integrated* Marketing *Communications: Building Brand Equity*. Pearson Education Limited, India.

Sihn, W. 2006. "A Review of Integrated Production Planning and Scheduling Systems." *Computers & Industrial Engineering* 49, no. 1: 346–359.

Slack, N., Chambers, S., Luo, T., and Johnson, S. 2016. *Operations Management*. 8th ed. Pearson Education Limited, India.

Solomon, M., Tassier, K. L., and Assaker, G. 2023. *Consumer Behavior: Conveying Marketing Strategy*. 13th ed. Pearson Education Limited, India.

Stevenson, M. L. 2009. *Operations Management in the Service Industry*. 8th ed. McGraw-Hill, New York.

Taj, S., Rahim, A., and Bani Hashim, S. H. 2011. "A Review of Production Analysis Models in Automotive Assembly Lines." *European Journal of Industrial Engineering* 5, no. 2: 137–153.

Teece, D. J. 2018. *Business Ecosystems: How Digital Platforms Are Changing the Way Business Is Done*. Harvard Business School Publishing, Cambridge.

Thomas, L. D. W., Audia, P. G., and Choudary, S. R. 2014. "Research Commentary—On Platforms, Ecosystems, and Strategic Decision-Making." *Strategic Management Journal* 35, no. 8: 1237–1241.

Trout, J., and Rivkin, R. 2009. *Differentiate or Die: Survival in Today's Marketplace*. Portfolio Penguin, India.

Verhoef, P. M. 2013. *Marketing in the Digital Age: Customer Focus and Transformation*. Kogan Page Publishers, New York.

Verhoef, P. M., Mick, D. G., and Bryant, J. 2003. "The Enduring Influence of the Marketing Mix on Contemporary Marketing Practice." *Journal of Marketing* 67, no. 3: 14–29.

Womack, J. P., Jones, D. T., and Roos, D. 1990. *The Machine That Changed the World*. Rawson Associates, England.

# 7 Modeling of Invisible Cost Factors in Information Technology Offshore Outsourcing by ISM Approach

*Prashant Chauhan, Nitin Kumar Chauhan, and Naveen J*

## 7.1 INTRODUCTION

The use of computerization started way back in the 1960s in the USA and other developed European countries. Due to lengthy computer programming at the starting stage of computer programming, software engineers tried to minimize coding for new programs and to represent the date; they mainly adopted a format as DD/MM/YY. In 1992, few organizations realized that in the year 2000, all such software failed to be distinguished between the year 1900 and the year 2000. At that time, enormous requirements for software developers arose due to simultaneous changes in all software, leading to Information Technology (IT) offshore outsourcing. Offshore outsourcing operations could be the production of goods or services that are outsourced to a different organization based in another country (Varadarajan 2009). After the 1990s, offshore operations reached the service sector worldwide (Stringfellow, Teagarden, en Nie 2008). Offshore outsourcing provides further advantages when conventional factors and law rules of the land are not adequate conditions for advancing economic progress. At the same time, however, the combination of human workforce with a subsequent additional factor of knowledge domain does appear to provide a much more incentive for industries to consider separating their core areas of working (Doh 2005; Saini 2020). Outsourcing may also help an organization get better, more advanced services than it could afford locally (Ellram, Tate, en Billington 2008). Offshore outsourcing of services may help build organizational resources; however, under certain situations, such outsourcing operations might lead to the depletion of critical human resources (Jensen 2012). Multiple factors are considered when deciding to outsource. By utilizing structured techniques, the best organizations may use informed choices along with their specific targets (Sharma et al. 2024). Lower labor

DOI: 10.1201/9781032700953-8

cost is the primary driving factor, but many organizations also aim to focus solely on their core businesses and create value in the market to ensure their own survival. Organizations even move towards essential operational and strategic functions rather than non-strategic ones (Pfannenstein en Tsai 2004). The difference between failed and successful outsourcing tasks may depend on competent functioning via well-designed contracts, whose section should specify service levels sustained during non-specified periods and limit expedient behavior between the service provider and client (Mao, Lee, en Deng 2008). Mittal (2020) also emphasized many risks related to manufacturing sectors (Ashwini Kumar, Mittal, en Goel 2023). Fraud and misinterpretation that may be detected by an outsourced organization are largely shown in health care insurance, and fuzzy logic may be applied a few times to detect it (Kalra et al. 2022). The coronavirus, also known as SARS-CoV-2, is a severe pandemic that has affected the globe and outsourcing businesses (Shrivastava et al. 2021). In offshore outsourcing, the risk of expedience may not be prevented due to physical distance but, in fact, delivered highly beneficial attributes and temporal distances (Chen en Lin 2019). Organizations use high-end technologies for risk analysis, and such technologies are used as outsources (Singh et al. 2024). Researchers did contemporary research on contextual relationships, emphasizing good technical competency (S. Kumar, Parveen, en Haleem 2014). Also, researchers recognized the importance of advanced IT techniques in the economic, political, and personal issues addressed (Rajendra et al. 2022; Ajay Kumar, Mittal, en Haleem 2022).

Research has found that offshore-outsourcing-related organizations face challenges in calculating invisible costs accompanied by inter-organizational relationships (Handley en Benton Jr 2013). The authors have used Interpretive Structural Modeling (ISM) in this research task, a robust methodology for recognizing inter-relationships among cost factors in IT offshore outsourcing.

The primary focus of this article is to determine the challenges concerned with invisible cost factors in IT offshore outsourcing, set up the relationship between these known cost factors using ISM, propose a structural model for invisible cost factors in IT offshore outsourcing, and classify the cost factors into different classes with the help of MICMAC (Cross-Impact Matrix Multiplication Applied to Classification) analysis.

The rest of this research work is arranged as follows: After a detailed introduction in part one, part two presents applied methodology. Part three presents the literature review regarding seven types of invisible cost factors. Part four presents the detailed technique of the ISM approach for modeling invisible cost factors in IT offshore outsourcing. A conclusion and further research avenues are presented in sections five and six.

## 7.2 AN OVERVIEW OF THE "INTERPRETIVE STRUCTURAL MODELING (ISM)" APPROACH

ISM is a learning process that organizes various interrelated factors into a comprehensive systematic model. These factors create complexity regarding the structure of the model, which may not be joined in a clear-cut manner. Dealing with a system where composition is not clearly defined may be challenging. Therefore, a methodological analysis should be formulated to help find a suitable structure related to a system, and ISM modeling accomplishes this. Many illustrations of the application of ISM have been presented in the literature review. Two fundamental concepts, transitivity and reachability, are essential to understand the ISM methodology.

Standard nomenclature to define relationships among cost factors is discussed as follows: The indication of relationship among the cost factors is denoted by four symbols as given below (*a* and *b*).

P: → cost factor *a* will reach cost factor *b*

Q: → cost factor *b* will reach cost factor *a*

R: → cost factors *a* and *b* will affect each other

S: → cost factors *a* and *b* will not influence each other

This data is shown as a table of 0 & 1, defined as the Initial Reachability Matrix. If cost factor *a* does not influence cost factor *b*, then the value in the table's cell (*a*, *b*) is 0, and if cost factor *a* influences cost factor *b*, then the entry in the cell (*a*, *b*) of the table is 1. If *a* influences *b* and *b* influences *c*, then due to transitivity, *a* also influences *c*.

## 7.3 IDENTIFICATION OF INVISIBLE COST FACTORS IN IT OFFSHORE OUTSOURCING

Identifying invisible cost factors in IT offshore outsourcing is made possible by studying failure in offshore outsourcing. There are many cases of opportunistic behavioral uncertainty, monitoring problems, inappropriate evaluation methods, and faulty operation mechanisms. Organizations that make a few additional efforts with respect to money and time might get more effective, and the best strategic actions might (Yeo en Saboori-Deilami 2017). Offshore service outsourcing organizations face several risks, which demand effective control (Cai, Ci, en Zou 2011). In inter-organizational relationships, coordination costs often show service-related and geographical-location-related factors as important outsourcing factors that may contribute to the complexity of operations (Handley en Benton Jr 2013). When supporting offshore-outsourcing services – such as IT, software development, and materials management – are outsourced, employee innovation will be reduced (Raiborn, Butler, en Massoud 2009). Risk related to transaction costs exists due to regulatory uncertainty in non-developed nations (Stratman 2008). Due to different historical contexts and political systems, political risk may be categorized as political risks, exchange rate risks, and sovereign risks (Aron, Clemons, en Reddi 2005). More dependency on service providers in offshore outsourcing always exhibits many invisible costs (Stringfellow, Teagarden, en Nie 2008). Cultural differences also impact the performance of outsourced services (Youngdahl, Ramaswamy, en Dash 2010). Due to long distances and destinations having different laws, languages, and cultures, many difficulties arise (Nakatsu en Iacovou 2009). The principal–agent relationship between the outsourcing provider and manufacturing service outsourcing enterprises is crucial for a high-quality product (Cai, Ci, en Zou 2011). The International Business research community has paid limited attention to offshore outsourcing of services (Ramamurti 2004). It shows that there is a necessity to understand better how to reduce risks in offshore-outsourcing-related projects, and a study is insisted

**TABLE 7.1**
**Invisible cost factors of IT offshore outsourcing**

| S.No. | Name of factors | Short name of factors |
|---|---|---|
| 1 | Relationship factor (Mao, Lee, en Deng 2008; Handley en Benton Jr 2013) | A |
| 2 | Political factor (Aron, Clemons, en Reddi 2005) | B |
| 3 | Factor due to cultural differences (Youngdahl, Ramaswamy, en Dash 2010; Nakatsu en Iacovou 2009) | C |
| 4 | Transaction factor (Stratman 2008) | D |
| 5 | Management factor (Stringfellow, Teagarden, en Nie 2008) | E |
| 6 | Strategy factor (Cai, Ci, en Zou 2011) | F |
| 7. | Opportunistic factor (Chen en Lin 2019) | G |

on understanding the role of various contracts and types of relationships among cost factors and their impact on the offshore success (Mathew en Chen 2013). Based on the literature review, cost factors related to IT organizations in offshore outsourcing are selected and shown in Table 7.1.

## 7.4 ISM APPROACH TO MODELING

### 7.4.1 Developing the Contextual Relationship among Factors

After selecting and listing the seven factors, the next task is to determine the relationships among various cost factors. An 'influence to' type of inter-relationship is utilized for this task, where one cost factor influences another. This creates a connection related to unseen cost factors.

Some decision makers of various offshore-outsourcing-related organizations were consulted to develop the relationships among the invisible cost factors. To establish a definite relationship among the factors and formulate the Structural Self-Interaction Matrix (SSIM), the following four alphabets describe the direction of the interaction among the cost factors (*a* and *b*).

1. P: → cost factor *a* will reach cost factor *b*
2. Q: → cost factor *b* will reach cost factor *a*
3. R: → cost factors *a* and *b* will help to alleviate each other
4. S: → cost factors *a* and *b* will not relate to each other

### 7.4.2 Development of a SSIM

Due to the inter-relationship among the cost factors, the SSIM was made and is shown in Table 7.2. The SSIM was changed into a matrix of 0 & 1, called the Final Reachability Matrix (FRM), by replacing P, Q, R, and S with 1 and 0 according to the condition. This transitivity concept is also considered, and Table 7.3 shows FRM with dependence level and driving level.

**TABLE 7.2**
**Structural self-interaction matrix**

| Factors | B | C | D | E | F | G |
|---|---|---|---|---|---|---|
| A | S | S | P | Q | Q | Q |
| B | | Q | P | S | S | Q |
| C | | | P | P | S | Q |
| D | | | | Q | Q | Q |
| E | | | | | Q | Q |
| F | | | | | | Q |
| G | | | | | | |

**TABLE 7.3**
**Final reachability matrix with dependence level and driving level**

| "Factors" | A | B | C | D | E | F | G | "Driving Power" |
|---|---|---|---|---|---|---|---|---|
| A | 1 | 0 | 0 | 1 | 0 | 0 | 0 | 2 |
| B | 0 | 1 | 0 | 1 | 0 | 0 | 0 | 2 |
| C | 0 | 1 | 1 | 1 | 1 | 0 | 0 | 4 |
| D | 0 | 0 | 0 | 1 | 0 | 0 | 0 | 1 |
| E | 1 | 0 | 0 | 1 | 1 | 0 | 0 | 3 |
| F | 1 | 0 | 0 | 1 | 1 | 1 | 0 | 4 |
| G | 1 | 1 | 1 | 1 | 1 | 1 | 1 | 7 |
| Dependence | 4 | 3 | 2 | 7 | 3 | 2 | 1 | |

**TABLE 7.4**
**Level identification process through iterations**

| Factors | Reachability | Antecedent | Intersection | Level |
|---|---|---|---|---|
| A | 1,4 | 1,5,6,7 | 1 | II |
| B | 2,4 | 2,3,7 | 2 | II |
| C | 2,3,4,5 | 3,7 | 3 | IV |
| D | 4 | 1,2,3,4,5,6,7 | 4 | I |
| E | 1,4,5 | 3,5,6,7 | 5 | III |
| F | 1,4,5,6 | 6,7 | 6 | IV |
| G | 1,2,3,4,5,6,7 | 7 | 7 | V |

### 7.4.3 Level-Identification Process Through Iterations

The cost factors for which the reachability and intersection remain the same are placed at the top level. Level identification for seven factors is completed in just five iterations and is presented in Table 7.4.

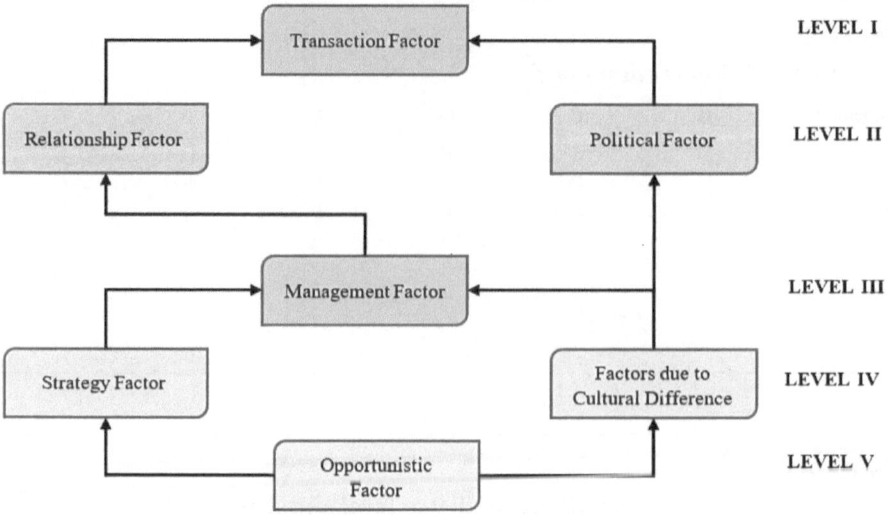


**FIGURE 7.1** Digraph showing the level of factors

### 7.4.4 Development of Digraph

*The development of the digraph* is drawn using lines of edges and nodes. The final digraph is drawn after removing the transitivity (Figure 7.1). If there is an interrelationship among the cost factors *a* and *b*, then it is shown with an arrow indicating from factor *a* to factor *b*.

### 7.4.5 MICMAC Analysis

The principle of MICMAC analysis is constructed on the properties of matrix multiplication. Its purpose is to recognize the cost factors' driving level and dependency level. This is done to identify the invisible cost factors related to IT offshore outsourcing. Based on these levels, the cost factors are divided into four categories as follows: autonomous factor, linkage factor, dependent factor, and independent factor. Figure 7.2 illustrates "driving level and dependence level diagram" as the results of the MICMAC analysis.

## 7.5 CONCLUSION

Based on the literature review, seven factors might affect invisible costs in offshore outsourcing operations. Further, an ISM model and MICMAC approach were simultaneously used. The results offer a critical classification of cost factors into four categories: i.e., (1) independent factor (opportunistic factor, factor due to cultural differences, and strategy factor). Opportunistic factor is a very dominating factor, further enhancing switching cost, low-level performance, and transaction costs. Factors due to cultural differences are very much dominating due to work habits and

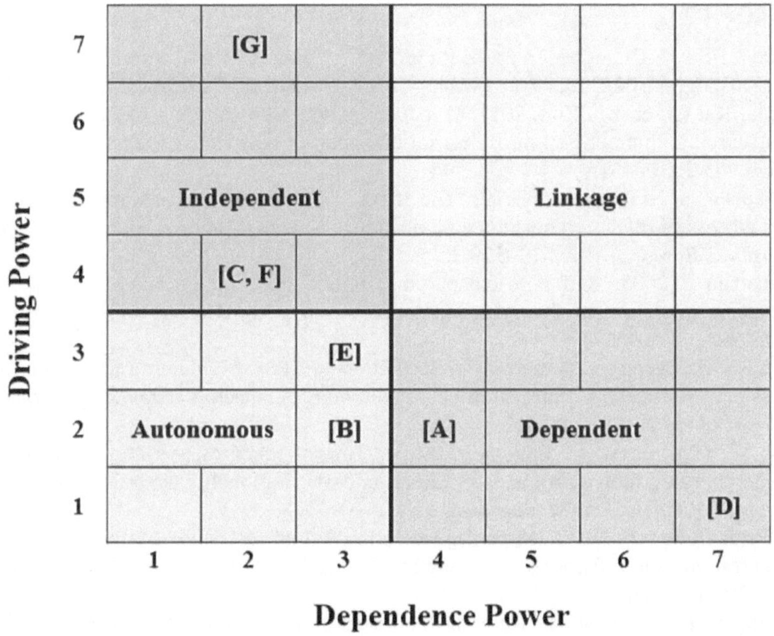


**FIGURE 7.2** Driving level and dependence level diagram

practices, sense of self, beliefs and attitudes, and communication language. Strategy factors depend on the reason for offshore outsourcing like expertise, scalability, and focus on core business only. (2) Linkage (no factor in this case), (3) dependent (relationship factor & transaction factor), and (4) autonomous (management factor & political factor). The results depict that with the help of the ISM approach details related to the driver and dependence level factors imbibed with invisible cost factors in offshore outsourcing could be got.

## 7.6 FUTURE SCOPE

Due to huge communication and advanced equipment developments, offshore outsourcing is now very much used. Real-time offshore control of operations is now readily available in many parts of the globe. In a few areas, there is still scope for further improvement. Issues of trust between organizations are very dominant factors in offshore outsourcing. Cultural differences related to work culture and different sex orientation acceptance are further needed to be adequately addressed. Many challenges are related to differences in human behavior due to changes in religious dogmas, environment, political situations, and compatibility for technical awareness. The social security issue is very dominant in South Asian countries for night shift working, which is essential due to the time-zone difference between the countries of client organizations as well as service provider organizations. Technical adaptability for advanced equipment and gadgets should be synchronous between both countries of business partners.

## REFERENCES

Aron, Ravi, Eric K. Clemons, en Sashi Reddi. 2005. "Just right outsourcing: Understanding and managing risk". *Journal of Management Information Systems* 22 (2): 37–55.

Cai, Sanfa, Kai Ci, en Bin Zou. 2011. "Producer services outsourcing risk control based on outsourcing contract design: Industrial engineering perspective". *Systems Engineering Procedia* 2: 308–315.

Chen, Stephen, en Nidthida Lin. 2019. "The effect of inter- and intra-organizational distances on success of offshored outsourced innovation: A configurational approach". *Journal of Business Research* 103: 519–529. https://doi.org/10.1016/j.jbusres.2018.03.023.

Doh, Jonathan P. 2005. "Offshore outsourcing: Implications for international business and strategic management theory and practice". *Journal of Management Studies* 42 (3): 695–704.

Ellram, Lisa M., Wendy L. Tate, en Corey Billington. 2008. "Offshore outsourcing of professional services: A transaction cost economics perspective". *Journal of Operations Management* 26 (2): 148–163.

Handley, Sean M., en W. C. Benton Jr. 2013. "The influence of task-and location-specific complexity on the control and coordination costs in global outsourcing relationships". *Journal of Operations Management* 31 (3): 109–128.

Jensen, Peter D. Ørberg. 2012. "A passage to India: A dual case study of activities, processes and resources in offshore outsourcing of advanced services". *Journal of World Business* 47 (2): 311–326.

Kalra, Gunjan, Yogendra Kumar Rajoria, Rahul Boadh, P Rajendra, Pratibha Pandey, Naveen Khatak, en Ajay Kumar. 2022. "Study of fuzzy expert systems towards prediction and detection of fraud case in health care insurance". *Materials Today: Proceedings* 56: 477–480. https://doi.org/10.1016/j.matpr.2022.02.157.

Kumar, Ajay, Ravi Kant Mittal, en Abid Haleem. 2022. *Advances in Additive Manufacturing: Artificial Intelligence, Nature-Inspired, and Biomanufacturing*. Elsevier.

Kumar, Ashwini, Ravi Kant Mittal, en Rajesh Goel. 2023. *Waste Recovery and Management: An Approach Toward Sustainable Development Goals*. CRC Press.

Kumar, S., R. Parveen, en A. Haleem. 2014. "Modeling of barriers to technology transfer implementation using system approach". *Udyog Pragati* 38 (2): 1–11.

Mao, Ji-Ye, Jae-Nam Lee, en Chun-Ping Deng. 2008. "Vendors' perspectives on trust and control in offshore information systems outsourcing". *Information & Management* 45 (7): 482–492.

Mathew, Saji K., en Yuanyuan Chen. 2013. "Achieving offshore software development success: An empirical analysis of risk mitigation through relational norms". *The Journal of Strategic Information Systems* 22 (4): 298–314.

Nakatsu, Robbie T., en Charalambos L Iacovou. 2009. "A comparative study of important risk factors involved in offshore and domestic outsourcing of software development projects: A two-panel Delphi study". *Information & Management* 46 (1): 57–68.

Pfannenstein, Laura L., en Ray J Tsai. 2004. "Offshore outsourcing: Current and future effects on American IT industry". *Information Systems Management* 21 (4): 72–80.

Raiborn, Cecily A., Janet B. Butler, en Marc F. Massoud. 2009. "Outsourcing support functions: Identifying and managing the good, the bad, and the ugly". *Business Horizons* 52 (4): 347–356.

Rajendra, P., Mina Kumari, Sangeeta Rani, Namrata Dogra, Rahul Boadh, Ajay Kumar, en Mamta Dahiya. 2022. "Impact of artificial intelligence on civilization: Future perspectives". *Materials Today: Proceedings* 56: 252–256.

Ramamurti, Ravi. 2004. "Developing countries and MNEs: Extending and enriching the research agenda". *Journal of International Business Studies* 35: 277–283.

Saini, C. P. 2020. Difficulty in transition: Study on hindrances in online shopping for potential shoppers. *International Journal of Online Marketing (IJOM)* 10 (2): 17–29.

Sharma, Parveen, Kashmir Singh Ghatorha, Amardeep Singh Kang, Lenka Cepova, Ajay Kumar, en Rakesh Kumar Phanden. 2024. "Strategic insights in manufacturing site selection: a multi-method approach using factor rating, analytic hierarchy process, and best worst method". *Frontiers in Mechanical Engineering* 10 (April): 1–11. https://doi.org/10.3389/fmech.2024.1392543.

Shrivastava, V. K., Ajay Kumar, A. Shrivastava, A. Tiwari, K. Thiru, en R. Batra. 2021. "Study and trend prediction of Covid-19 cases in India using deep learning techniques". In *Journal of Physics: Conference Series*, 1950:12084. IOP Publishing.

Singh, Pooja, Lipsa Das, Raju Jha, Ajay Kumar, en Sangeeta Rani. 2024. "Revolutionizing Sales and Marketing: The Power of AI in Action". In A. Kumar, S. Rani, S. Rathee, & S. Bhatia (Eds.), *Security and Risk Analysis for Intelligent Cloud Computing*, 297–315. CRC Press.

Stratman, Jeff K. 2008. "Facilitating offshoring with enterprise technologies: Reducing operational friction in the governance and production of services". *Journal of Operations Management* 26 (2): 275–287.

Stringfellow, Anne, Mary B. Teagarden, en Winter Nie. 2008. "Invisible costs in offshoring services work". *Journal of Operations Management* 26 (2): 164–179.

Varadarajan, Rajan. 2009. "Outsourcing: Think more expansively". *Journal of Business Research* 62 (11): 1165–1172.

Yeo, ChangSeob, en Vafa Saboori-Deilami. 2017. "Strategic challenges of outsourcing innovation in global market". *Asia Pacific Journal of Innovation and Entrepreneurship* 11 (1): 5–16.

Youngdahl, William E., Kannan Ramaswamy, en Kishore C. Dash. 2010. "Service offshoring: the evolution of offshore operations". *International Journal of Operations & Production Management* 30 (8): 798–820.

# 8 Analysis of Dynamics between Cryptocurrencies and Foreign Exchange Rates in Indian Scenario

*Rohan Bhargava and Hem Shweta Rathore*

## 8.1 INTRODUCTION

The advent and growth of cryptocurrencies have brought about a revolutionary change in the financial world by integrating technology and deep socio-economic implications. The origin of cryptocurrencies, specifically Bitcoin, marked the birth of a Decentralized Finance (DeFi) system that has been developed by scholars who argue that blockchain is necessary to ensure this decentralization [1]. Cryptocurrencies' functionality and potential are mostly driven by technological developments in blockchain [2]. This perspective is similarly echoed through a discussion on how blockchain remains foundational for various applications of cryptocurrency [3, 4]. This refers to the use of technology as a basis for security, transparency, and innovation in financial services. Likewise, cryptocurrency versus darknet market relationship has been explored, revealing that these digital assets could also be used illicitly [5]. The volatility nature and dynamics of markets surrounding cryptocurrencies have been likened to speculative bubbles depicting them as similar to moderately sized stock exchanges [6]. Furthermore, the volatility spillover effects and the interconnectedness of cryptocurrencies with broader financial markets suggest a complex relationship that impacts global economic stability [7]. The discourse extends into the socio-economic and sustainability concerns associated with cryptocurrencies, pointing to the potential for economic disruption alongside environmental considerations [8].

Globally, cryptocurrencies have challenged traditional financial institutions and regulatory frameworks, leading to a dynamic interplay between innovation, regulation, and market adaptation. India's relationship with cryptocurrency provides a compelling case study of this dynamic. The Indian market has shown a keen interest in cryptocurrencies, evidenced by the rapid growth of crypto exchanges and the increasing adoption of digital currencies among its population.

 DOI: 10.1201/9781032700953-9

Despite initial regulatory challenges, including the Reserve Bank of India's (2018) directive against cryptocurrency transactions, the 2020 Supreme Court ruling in favor of lifting these restrictions marked a pivotal turn, fostering a more favorable environment for cryptocurrency ventures and innovation in India. The Indian government continues to proceed with caution when it comes to cryptocurrencies. They have expressed concerns regarding money laundering, terrorism financing and consumer protection, and have indicated that the government is "working on" a legislation to regulate cryptocurrencies in the country to address these issues and provide clarity in the system [9, 10].

In the case of cryptocurrencies, regulatory and ethical issues emerge as significant concerns because they have decentralized nature without centralized control. Challenges and opportunities arise from the integration of cryptocurrencies with traditional financial systems, which highlights the need for adaptive regulation policies to connect digital currencies and fiat ones [11]. This is also seen in the way digital currencies are being used in global finance [12]. Regulations are needed urgently to make sure that there are ethical practices in cryptocurrency transactions [13]. The Bitcoin community's underlying moral values provide for a mosaic of ethical and regulatory perspectives that shape market trends on cryptocurrencies [14]. In this regard, looking at its history, it can be said that cryptocurrency has been serving both as an investment asset and medium of exchange within the digital economy; this duality exposes regulatory hurdles connected with cryptos such as their use for illegal purposes besides their intrinsically high volatility rates [15].

The integration of cryptocurrencies into existing tax systems presents significant challenges, particularly due to their pseudonymity and dual function as both assets and currency. Corrective taxation is essential to address carbon emissions from mining, and there is significant potential for capital gains tax revenue from cryptocurrencies. The concentration of cryptocurrency ownership and challenges for VAT/sales taxes necessitate adaptations in tax systems to manage cryptocurrencies effectively. While focusing on income tax implications, there is also a critical need to explore sales taxation, as cryptocurrencies pose profound risks to VAT systems, especially in emerging economies with strong crypto demand and weaker tax administration [16].

The rise of DeFi, an offshoot of cryptocurrency technology, has attracted more attention, and a broad overview is provided on how blockchain and smart contracts have allowed new types of financial services to be developed that can operate independently from traditional bank systems [17]. This investigation into DeFi highlights the potential for cryptocurrencies in transforming the activities of financial institutions emphasizing the necessity to understand both the risks and opportunities offered by this technology. Furthermore, decentralized exchange and lending protocols have been underscored as a core component in DeFi. Therefore, through a novel algorithm and analysis of protocol interactions, we examine nested building blocks within DeFi compositions providing insights into its complexity in relation to traditional finance. Additionally, the study identifies possible systemic risks owing to DeFi's rapid growth as well as complexity, which also provides a greater understanding for disruption capabilities; hence, informed regulatory and developmental strategies are needed for going forward [18].

## 8.2 LITERATURE REVIEW

The academic investigation concerning cryptocurrencies has greatly expanded, touching various perspectives from barriers of adoption to market dynamics, legal frameworks, and wider implications on conventional financial systems. In the early stages, research critically analyzed Bitcoin's adoption in India and noted that the volatility of the market and regulatory gaps were key hindrances to incorporating cryptocurrencies into emerging economies such as India [19, 20]. Also, consistent with this narrative is the lack of a well-defined legal framework for cryptocurrencies in India, which further complicates efforts to ensure market stability and security against fraud [21]. Consequently, calls have been made for a more organized manner of threading through digital currencies into India's financial fabric [22].

The conversation then shifted towards unveiling herding behavior in cryptocurrency markets, implying a great deal about investor psychology compared to traditional financial assets and highlighting the peculiarities of these markets along with their disruptive capacity [23]. Meanwhile, an examination of how global events impact foreign exchange rates through looking at cryptocurrencies exposed their interdependence across borders during times like COVID-19 pandemic period [24]. At the same time, blockchain has been transforming the forex sector while offering hints regarding possible challenges associated with regulation and vulnerability to attack [25].

Comparing its volatility with traditional forex market gives insights on potential investment strategies in cryptocurrencies [26]. This idea of volatility was also mentioned before and laid down foundation for understanding dynamic linkages between cryptos and fiat currencies, especially emphasizing on Bitcoin's high correlation with United States Dollar (USD) [27, 28, 29]. The study analyses show how cryptocurrencies are now increasingly being used as hedging instruments in forex markets. First, studies reveal how major cryptocurrencies—Bitcoin, Ethereum, Litecoin and Ripple—can be used as hedge tools against foreign currency risks [30]. This research demonstrated how various forms of electronic cash present better ways to hedge against volatile prices than gold or an assorted set of currencies because each has an independent value making them unique among other factors such as price variation among different products sold internationally as well as factors affecting their price in comparison with other traditional banks [30].

To build on this, the research explored how cryptocurrencies are cross-correlated to traditional currencies such as Euro (EUR) and USD, suggesting that cryptocurrency markets may be decoupling from forex [31]. This observation demonstrates that cryptos can be considered as safe havens or alternative hedging mechanisms in response to traditional currency market fluctuations. Moreover, the deep interconnections and mutual influences across digital and traditional currency landscapes have been underscored by further analysis of dynamic spillovers and causality between major cryptocurrencies and leading currency markets, with significant spillover effects especially from Bitcoin and Litecoin to various global currencies [32].

Financial investigations into Bitcoin's properties were centered on its interaction with financial indicators using different multivariate Generalized Autoregressive Conditional Heteroskedasticity (GARCH) models. The research examines the diversification and hedging effectiveness of Bitcoin vis-à-vis gold assets and stock market, resulting in substantial returns and volatility spillovers [33]. The analysis recognizes VARMA (1, 1)-DCC-GJR-GARCH model as the most effective in terms of modeling financial assets' dynamics, indicating that short positions against Bitcoins can perfectly hedge investment risks within varied financial assets. On an additional note, including Bitcoin in portfolios alongside gold, oil as well as equities helps to reduce overall risk, thereby justifying its role in enhancing portfolio risk management [33].

In addition to that it was also possible for detailed comparison study to show how COVID-19 pandemic has affected weak-form efficiency of cryptocurrency and forex markets. The study thoroughly analyzed these markets' efficiency over time using overlapping sliding windows employing asymmetry multifractal detrended fluctuation analysis method through asymmetric market deficiency measure, fuzzy entropy together with Tsallis entropy and Fisher information for each market. The analysis identifies a number of significant shifts in market efficiency during this period compared with pre-pandemic times particularly showing an increase in weak-form inefficiency among traditional currencies during this period. In fact, behavior of Bitcoin was similar to that of regular currencies, thereby confirming its uniqueness within the cryptocurrencies [34]. The study's insights on pandemic impact on market efficiency under various market conditions are very useful for investors, portfolio managers and policy makers.

Further research has explored the complex dynamics between Bitcoin and major global currencies, emphasizing Bitcoin's increasing integration with returns on major currencies such as EUR and Chinese Yuan [35, 36]. This integration was further examined revealing how Bitcoin affects traditional financial indices as well as how it is not attached to key global currencies at all [37, 38].

The understanding of how Bitcoin influences traditional exchange rates and its potential in stimulating economic development was greatly enriched through collaborative studies [39, 40, 41]. These papers, as well as other ones that deepen our knowledge about the interaction between cryptocurrencies and various financial markets, show the complicated multilateral relations that exist between digital currencies and traditional financial systems [42, 43].

This vast amount of research not only explains the complexities and problems that are present within the cryptocurrency world but also showcases how digital currencies can revolutionize traditional financial transactions. They demonstrate that flexible regulations, advanced technologies and a prudential strategy to exploit chances while curbing hazards emanating from virtual coins will remain topical concerns for years to come.

## 8.3 RESEARCH METHODOLOGY

The aim of the study is to look at how foreign currency exchange affects cryptocurrencies, the relationship between foreign currency exchange and cryptocurrencies and evaluate the existence of autocorrelation in residuals from regression models predicting cryptocurrency prices using foreign exchange. This research uses a descriptive research design as it was best suited for this type of investigation by relying on secondary data only. The main goal in this paper is to offer a brief overview about digital currencies with a focus on the major ones such as Bitcoin, Ethereum and Litecoin. Using purposive sampling technique, this study will select Bitcoin, Ethereum and Litecoin, while USD, Great British Pound (GBP) and EUR were selected for testing their relationships with Bitcoin—all these are major forex rates. To help correct for some of the limitations inherent in government-regulated data sources, we chose investing.com which is known for its deep coverage of global financial markets. It was selected because it provides reliable data that spans over time making it ideal for carrying out research. By analyzing historical data, the report aims at examining how cryptocurrency valuations have related to fluctuating forex rates over time. In order to conduct research analysis, correlation, regression and Durbin–Watson tests were applied to the collected data.

### 8.3.1 Research Objectives

1. To study the impact of selected foreign currency exchange on selected cryptocurrencies.
2. To find the correlation between selected foreign currency exchange and selected cryptocurrencies.
3. To assess the independence of error terms in a regression model that predicts cryptocurrency prices using foreign exchange rates.

## 8.4 DATA ANALYSIS

### 8.4.1 Descriptive Statistics

The descriptive statistics for Bitcoin, Ethereum and Litecoin, and the foreign exchange rates of USD, GBP and EUR provide us with a comprehensive overview of the quantitative data for each variable (Table 8.1).

Bitcoin has a wide range between its minimum and maximum values, suggesting high volatility in its price during the period under study. Ethereum and Litecoin also show significant ranges but are less volatile compared to Bitcoin. The foreign exchange rates are relatively stable as indicated by their narrower ranges.

The quartiles for each cryptocurrency show a skewed distribution, with the median being lower than the mean, especially for Bitcoin. This indicates a right-skewed distribution, typical for asset prices. For foreign exchange rates, the median and mean are relatively close, suggesting a more symmetric distribution.

Bitcoin has the highest mean value, indicating it is the most valuable of the three cryptocurrencies on average. Ethereum and Litecoin have lower mean values, reflecting their lower market prices compared to Bitcoin. The mean values of the foreign exchange rates are within expected ranges, showing the average exchange rate during the observed period.

Bitcoin has a very high variance and standard deviation, indicating significant price fluctuation or volatility over the period studied. Ethereum has a lower variance and standard deviation compared to Bitcoin, but it still indicates considerable volatility. Litecoin has the lowest variance and standard deviation among the

**TABLE 8.1**
**Descriptive statistics**

| Statistic | Bitcoin | Ethereum | Litecoin | USD price | GBP price | EUR price |
|---|---|---|---|---|---|---|
| Number of observations | 1303 | 1303 | 1303 | 1303 | 1303 | 1303 |
| Minimum | 3401.376 | 104.645 | 30.378 | 68.410 | 83.640 | 76.256 |
| Maximum | 67549.734 | 4810.071 | 387.869 | 83.426 | 107.743 | 92.446 |
| 1st Quartile | 9665.775 | 232.393 | 56.538 | 72.718 | 93.134 | 80.538 |
| Median | 22156.406 | 1445.151 | 75.495 | 74.750 | 98.344 | 85.390 |
| 3rd Quartile | 36727.959 | 1974.662 | 114.880 | 81.101 | 101.478 | 88.260 |
| Mean | 24550.708 | 1407.950 | 94.542 | 75.923 | 97.263 | 84.548 |
| Variance ($n$-1) | 258496276.139 | 1366182.743 | 3056.904 | 20.156 | 27.856 | 18.013 |
| Standard deviation ($n$-1) | 16077.819 | 1168.838 | 55.289 | 4.490 | 5.278 | 4.244 |

*Source:* Author's own calculation
*Abbreviations:* EUR, Euro; GBP, Great British Pound; USD, United States Dollar.

cryptocurrencies, suggesting it is the least volatile. The foreign exchange rates have much lower variances and standard deviations, indicating their stability compared to the highly volatile cryptocurrency market.

The descriptive statistics reflect the inherent volatility in cryptocurrency prices, with Bitcoin as the most volatile and valuable among the three. Ethereum and Litecoin follow, with Ethereum being more volatile and valuable than Litecoin.

- The distribution of prices for all three cryptocurrencies suggests that there are periods of significant price increases, which is consistent with the known price behavior of cryptocurrencies that can undergo large price swings in short periods.
- The foreign exchange rates are much more stable, as expected from regulated and heavily traded markets. They do not exhibit the same level of volatility as cryptocurrencies.
- The comparative analysis between cryptocurrencies and forex rates would also highlight the difference in risk and potential returns, where cryptocurrencies may offer higher potential returns at a significantly higher risk, as indicated by their standard deviations.

## 8.4.2 Correlation

### 8.4.2.1 Correlation Matrix (Pearson)

Bitcoin, Ethereum and Litecoin show strong positive correlations with each other (Bitcoin & Ethereum: 0.920, Bitcoin & Litecoin: 0.814, Ethereum & Litecoin: 0.703) (Table 8.2). This indicates that as the price of one of these cryptocurrencies goes up or down, the others are likely to move in the same direction. USD price has a positive but weak correlation with Bitcoin (0.247) and Ethereum (0.337), and a negative

**TABLE 8.2**
**Correlation between selected cryptocurrencies and forex rates**

| Variables | Bitcoin | Ethereum | Litecoin | USD price | GBP price | EUR price |
|---|---|---|---|---|---|---|
| Bitcoin | 1 | 0.920 | 0.814 | 0.247 | 0.730 | 0.549 |
| Ethereum | 0.920 | 1 | 0.703 | 0.337 | 0.693 | 0.465 |
| Litecoin | 0.814 | 0.703 | 1 | –0.160 | 0.455 | 0.345 |
| USD price | 0.247 | 0.337 | –0.160 | 1 | 0.574 | 0.566 |
| GBP price | 0.730 | 0.693 | 0.455 | 0.574 | 1 | 0.893 |
| EUR price | 0.549 | 0.465 | 0.345 | 0.566 | 0.893 | 1 |

*Source:* Author's own calculation

**TABLE 8.3**
**Pearson's coefficients between selected cryptocurrencies and forex rates**

| Variables | Bitcoin | Ethereum | Litecoin | USD price | GBP price | EUR price |
|---|---|---|---|---|---|---|
| Bitcoin | 1 | 0.846 | 0.662 | 0.061 | 0.533 | 0.301 |
| Ethereum | 0.846 | 1 | 0.494 | 0.113 | 0.480 | 0.217 |
| Litecoin | 0.662 | 0.494 | 1 | 0.026 | 0.207 | 0.119 |
| USD price | 0.061 | 0.113 | 0.026 | 1 | 0.329 | 0.320 |
| GBP price | 0.533 | 0.480 | 0.207 | 0.329 | 1 | 0.798 |
| EUR price | 0.301 | 0.217 | 0.119 | 0.320 | 0.798 | 1 |

*Source:* Author's own calculation

weak correlation with Litecoin (-0.160). This suggests a less-consistent relationship between USD exchange rates and cryptocurrency prices. GBP price shows a strong positive correlation with Bitcoin (0.730) and Ethereum (0.693), and a moderate correlation with Litecoin (0.455). This indicates that GBP exchange rate movements are more consistently associated with movements in cryptocurrency prices. EUR price has moderate positive correlations with Bitcoin (0.549) and Ethereum (0.465), and a weaker one with Litecoin (0.345), suggesting some level of association.

### 8.4.2.2 Coefficients of Determination (Pearson)

The coefficient of determination, $R^2$, is the square of the Pearson correlation coefficient and indicates the proportion of the variance in the dependent variable that is predictable from the independent variable (Table 8.3).

Bitcoin & Ethereum ($R^2 = 0.846$): This very high $R^2$ value suggests that approximately 84.6% of the variance in Bitcoin prices can be explained by the variance in Ethereum prices, indicating a very strong relationship.

Bitcoin & Litecoin ($R^2 = 0.662$): A substantial proportion (66.2%) of the variance in Bitcoin prices can be explained by the variance in Litecoin prices.

Cryptocurrencies seem to move together, with strong intercorrelations, suggesting that market sentiment and global factors affecting one major cryptocurrency are likely to affect others.

The foreign exchange rates have moderate-to-strong correlations with cryptocurrencies, but the strength of the relationship varies by currency and cryptocurrency. GBP shows a consistent influence on all three cryptocurrencies, whereas USD's influence is less consistent, and EUR lies somewhere in between. The relatively weaker correlations of USD price with cryptocurrencies could be attributed to the global status of USD as a reserve currency, which might be influenced by factors that do not directly affect cryptocurrencies. The strong correlation between GBP and EUR prices is expected, as they are both major currencies in closely linked economies.

While there are varying degrees of association, the analysis supports the idea that foreign exchange rates do have a relationship with cryptocurrency prices. This relationship appears to be stronger for some currency pairs (like GBP with Bitcoin and Ethereum) and weaker for others (like USD with Litecoin). The high intercorrelation between the cryptocurrencies themselves suggests a shared market dynamic or investor sentiment driving their prices.

### 8.4.3 Durbin–Watson Statistics

The study meticulously explores the relationship between foreign exchange rates and cryptocurrency prices, employing correlation coefficients and the Durbin–Watson statistics for rigorous quantitative analysis (Table 8.4). Key findings reveal the following:

Bitcoin exhibits a strong correlation with GBP and a moderate one with EUR, while its correlation with USD is weak. Ethereum demonstrates a strong positive correlation with GBP and a moderate negative one with EUR. Litecoin shows a strong positive correlation with GBP and a moderate negative correlation with USD.

These statistical correlations highlight the significant impact of forex rate fluctuations on cryptocurrency valuations. Moreover, the detection of positive autocorrelation across all examined cryptocurrencies suggests the intricacies of their integration into the global financial market and the necessity for revised monetary policies and regulatory frameworks.

**TABLE 8.4**
**Durbin–Watson statistics**

| Cryptocurrency | DW statistic | Interpretation |
|---|---|---|
| Bitcoin | 0.044 | Positive autocorrelation |
| Ethereum | 0.045 | Positive autocorrelation |
| Litecoin | 0.053 | Positive autocorrelation |

*Source:* Author's own calculation

The study's alignment with existing literature on the transformative role of blockchain technology and the speculative nature of cryptocurrencies further underscores the need for advanced econometric approaches and comprehensive research into the broader economic implications of these findings. The pronounced effects of GBP and EUR on cryptocurrency valuations not only mark critical areas for future investigation but also have profound implications for investment strategies, monetary policy formulation and the sustainability of the digital economy. The quantitative evidence provided, such as the strong correlation of Bitcoin with GBP (0.730) and the positive autocorrelation indicated by the Durbin–Watson statistics (Bitcoin: 0.044, Ethereum: 0.045, Litecoin: 0.053), reinforces the study's conclusions and underscores the importance of statistical analysis in deciphering the complex dynamics of financial markets.

## 8.5 DISCUSSION

The study reveals a complex relationship between foreign exchange rates and cryptocurrencies like Bitcoin, Ethereum and Litecoin, demonstrating significant impacts of GBP and EUR on cryptocurrency valuations. It evidences that fluctuations in forex rates correlate with changes in cryptocurrency prices, highlighting digital currencies' growing integration into the global financial market and prompting a re-evaluation of traditional monetary policies and regulatory frameworks.

Regression analysis reveals a strong positive correlation between cryptocurrencies and GBP, and nuanced relationships with USD and EUR, offering key insights for investors and policymakers. It also points to potential autocorrelation issues, suggesting future research avenues including the adoption of advanced econometric techniques, the use of longitudinal data and the examination of emerging technologies and geopolitical events' effects.

The novelty of this research lies in its comprehensive approach to analyzing the interplay between traditional currencies and cryptocurrencies within an emerging market context, a combination that has been less explored in existing literature. This study provides empirical evidence highlighting the intricate relationships between these financial instruments and emphasizes the need for dynamic regulatory frameworks alongside sophisticated econometric approaches to capture the evolving financial landscape.

## 8.6 CONCLUSION

The Indian cryptocurrency market was studied, focusing on how it is related to the forex rates. From analyzing the effects and relationships of certain benchmarks for selected forex-cryptocurrencies with a case study of cryptocurrencies' prices and testing for autocorrelation in regression models, important insights have been generated.

Our findings reveal that there were distinctive patterns in the connection between cryptocurrency prices and currency exchange rates. In particular, we found an upward linear relationship as well as predictive capacity of GBP on digital currency value whereas such links did not exist regarding USD or EUR. The Durbin–Watson statistics confirmed that our regression models were not suffering from autocorrelation, thus giving our results more credibility.

This research is distinct due to its all-inclusive coverage of traditional currencies versus cryptocurrency within the context of emerging markets. By presenting empirical evidence about these financial instruments' complex relationships, this paper contributes to the existing literature by emphasizing on the significance of flexible regulation frameworks and advanced econometric tools required to effectively deal with the dynamic nature of the financial environment.

The research provides considerable insight into how traditional currencies and cryptocurrencies are related. Notably, unlike USD or EUR, GBP has been shown to be highly predictive when it comes to evaluating cryptocurrencies. These findings should help policy makers and finance practitioners understand complexity in trading systems while also underlining the need for adaptive regulations.

Further studies are recommended to explore long-term implications of these relationships on market stability and regulatory practices. Future research should use longitudinal designs incorporating new econometric techniques in order to fully comprehend and forecast dynamics between conventional monetary system vis-à-vis digital currencies. This can keep pace with ongoing financial innovations necessitating responsive regulatory frameworks as well as marketing strategies.

### 8.6.1 Future Scope of Study

The study faces limitations, notably its reliance on secondary data that may impact the generalizability of findings. Future research could enhance understanding by incorporating longitudinal studies, employing more sophisticated statistical methods and exploring the impact of regulatory changes and technological advancements on the cryptocurrency–forex relationship.

In light of these findings, investors are advised to consider forex market trends in their cryptocurrency investment strategies. Policymakers should strive for adaptable regulatory frameworks that recognize digital currencies' unique attributes, ensuring market stability and investor protection. Businesses and financial institutions could also benefit from the insights provided, developing strategies that navigate the volatility of cryptocurrencies in relation to traditional foreign exchange rates.

Contributing to the financial research field, this study underscores the intricate relationship between foreign exchange rates and cryptocurrencies, advocating for a nuanced approach to digital currencies' role within the financial ecosystem. This study provides empirical evidence highlighting the intricate relationships between these financial instruments and emphasizes the need for dynamic regulatory frameworks alongside sophisticated econometric approaches to capture the evolving financial landscape. It calls for continued research, adaptive regulatory measures and strategic investment practices to navigate the evolving digital financial landscape.

### 8.6.2 Implications

The implications of the study highlight the significant correlation between foreign exchange rates and cryptocurrency valuations, affecting investors, policymakers, businesses and the academic community. Investors can use this knowledge to refine their trading strategies, enhancing portfolio diversification and optimizing returns by understanding the impact of traditional currency fluctuations on digital assets. Policymakers are prompted to develop adaptable regulatory frameworks that account for digital currencies' distinct features, aiming for market stability and innovation. Businesses and financial institutions integrating cryptocurrencies can benefit from improved risk management and strategic planning by acknowledging currency risks. The academic sector is encouraged to further investigate the dynamics between foreign exchange rates and cryptocurrencies, exploring their broader economic and environmental implications. The study's insights suggest a comprehensive approach for stakeholders to navigate the evolving digital financial ecosystem, emphasizing the need for informed investment decisions, regulatory flexibility and ongoing research.

## 8.7 DECLARATION

This article is an original work of authors and the same has not been published earlier in any other publication.

There is no conflict of interest.

## 8.8 ETHICAL CONSIDERATIONS

Given the use of secondary data, this study ensures that all data sources are credited and that the data is used responsibly, adhering to relevant data protection regulations and ethical guidelines.

## REFERENCES


1. Vujičić, Dejan, Dijana Jagodić, and Siniša Ranđić. "Blockchain technology, bitcoin, and Ethereum: A brief overview." In *2018 17th International Symposium Infoteh-Jahorina (Infoteh)*, pp. 1–6. IEEE, 2018.
2. Chaudhari, Mr Yash D. "Cryptocurrency market analysis: Trends and future outlook." In *International Conference on Enhancing Productivity in Hybrid Mode: The Beginning of a New Era*: 331. ISBN: 978-81-950596-5-2.
3. Kumar, G. P. "Cryptocurrency–the next big thing." *Recent Trends in Management and Commerce* 3, no. 1 (2022): 14–19.
4. Amsyar, Izwan, Ethan Christopher, Arusyi Dithi, Amar Najiv Khan, and Sabda Maulana. "The challenge of cryptocurrency in the era of the digital revolution: A review of systematic literature." *Aptisi Transactions on Technopreneurship (ATT)* 2, no. 2 (2020): 153–159.
5. Janze, Christian. (2017). "Are cryptocurrencies criminals best friends? Examining the co-evolution of bitcoin and darknet markets." In *23rd Americas Conference on Information Systems (AMCIS 2017)*. ISBN 978-0-9966831-4-2.
6. Wątorek, Marcin, Stanisław Drożdż, Jarosław Kwapień, Ludovico Minati, Paweł Oświęcimka, and Marek Stanuszek. "Multiscale characteristics of the emerging global cryptocurrency market." *Physics Reports* 901 (2021): 1–82.
7. Liu, Jinan, and Apostolos Serletis. "Volatility in the cryptocurrency market." *Open Economies Review* 30, no. 4 (2019): 779–811.
8. Giudici, Giancarlo, Alistair Milne, and Dmitri Vinogradov. "Cryptocurrencies: Market analysis and perspectives." *Journal of Industrial and Business Economics* 47 (2020): 1–18.
9. https://www.rbi.org.in/commonman/Upload/English/PressRelease/PDFs/PR264205042018.PDF

10. https://main.sci.gov.in/supremecourt/2018/19230/19230_2018_4_1501_21151_Judgement_04-Mar-2020.pdf
11. Rejeb, Abderahman, Karim Rejeb, and John G. Keogh. "Cryptocurrencies in modern finance: A literature review." *Etikonomi* 20, no. 1 (2021): 93–118.
12. García-Corral, Francisco Javier, José Antonio Cordero-García, Jaime de Pablo-Valenciano, and Juan Uribe-Toril. "A bibliometric review of cryptocurrencies: How have they grown?" *Financial Innovation* 8 (2022): 1–31.
13. Owolabi, S. A., and Tunde Osanyinbi. "The historical evolution, developments and accounting implications of cryptocurrency operations." *European Journal of Accounting, Auditing and Finance Research* 11, no. 10 (2023): 55–72.
14. Tana, Silviana, and Christoph F. Breidbach. "Institutionalizing digital transformation through cryptocurrency use." In *29th European Conference on Information Systems (ECIS)*, p. 107, 2021. ISBN 978-1-7336325-6-0.
15. Nam, Nguyen Hoang. "Impact of cryptocurrencies on financial markets." *The VMOST Journal of Social Sciences and Humanities* 65, no. 2 (2023): 03–15.
16. Baer, Katherine, Ruud De Mooij, Shafik Hebous, and Michael Keen. "Taxing cryptocurrencies." *Oxford Review of Economic Policy* 39, no. 3 (2023): 478–497.
17. Li, Ye, and Simon Mayer. "Money creation in decentralized finance: A dynamic model of stablecoin and crypto shadow banking." *Fisher College of Business Working Paper* 2020–03 (2022): 030.
18. Auer, Raphael, Bernhard Haslhofer, Stefan Kitzler, Pietro Saggese, and Friedhelm Victor. "The technology of decentralized finance (DeFi)." *Digital Finance* 6, no. 1 (2024): 55–95.
19. Yousuf Javed, Mohd, Mohammad Hasan, and Ruzaina Khan. "Future of bitcoin in India: Issues and challenges." *Journal of Statistics and Management Systems* 23, no. 2 (2020): 207–214.
20. Banwari, Vijeta. "Cryptocurrency scope in India." *International Research Journal of Management Sociology & Humanities* 8, no. 12 (2017): 82–92.
21. Vishwakarma, Paras, Z. Khan, and Taruna Jain. "Cryptocurrency, security issues and upcoming challenges to legal framework in India." *International Research Journal of Engineering and Technology* 5, no. 1 (2018): 212–215.
22. Thakur, Deshant Singh, Raj A. Varma, and Damodar Mayappa Hake. "Regulation of cryptocurrency in India: Issues and challenges." *Journal of Positive School Psychology* 6, no. 5 (2022): 8921–8929.
23. Gemayel, Roland, and Alex Preda. "Herding in the cryptocurrency market: A transaction-level analysis." *Journal of International Financial Markets, Institutions and Money* 91 (2024): 101907.
24. Almansour, Bashar Yaser, Md Mohan Uddin, Sabri Elkrghli, and Ammar Yaser Almansour. "The dynamic connectedness between cryptocurrencies and foreign exchange rates: evidence by TVP-VAR approach." *Industrial Engineering & Management Systems* 22, no. 3 (2023): 349–362.
25. Moşteanu, Narcisa Roxana, and Tatiana Flocea. "Revolutionizing foreign exchange market: A critical analysis of blockchain's opportunities and challenges." *European Journal of Theoretical and Applied Sciences* 1, no. 5 (2023): 33–44.
26. Chinthapalli, Usha Rekha. "A comparative analysis on probability of volatility clusters on cryptocurrencies, and FOREX currencies." *Journal of Risk and Financial Management* 14, no. 7 (2021): 308.
27. Corelli, Angelo. "Cryptocurrencies and exchange rates: A relationship and causality analysis." *Risks* 6, no. 4 (2018): 111.
28. Kuzucu, Seyfettin Caner. "Investigation of the relationship between bitcoin and exchange rates." In *3rd International Conference on Economic Research*, pp. 156–160. 2019.
29. Smith, Jacob. "An analysis of Bitcoin exchange rates." *Available at SSRN 2493797* (2016).

30. Cheong, Calvin W. H. "Cryptocurrencies vs global foreign exchange risk." *The Journal of Risk Finance* 20, no. 4 (2019): 330–351.
31. Drożdż, Stanisław, Ludovico Minati, Paweł Oświęcimka, Marek Stanuszek, and Marcin Wątorek. "Signatures of the crypto-currency market decoupling from the Forex." *Future Internet* 11, no. 7 (2019): 154.
32. Elsayed, Ahmed H., Giray Gozgor, and Chi Keung Marco Lau. "Causality and dynamic spillovers among cryptocurrencies and currency markets." *International Journal of Finance & Economics* 27, no. 2 (2022): 2026–2040.
33. Saini, C. P. "Difficulty in transition: Study on hindrances in online shopping for potential shoppers." *International Journal of Online Marketing (IJOM)* 10, no. 2 (2020): 17–29.
34. Zitis, Pavlos I., Shinji Kakinaka, Ken Umeno, Michael P. Hanias, Stavros G. Stavrinides, and Stelios M. Potirakis. "Investigating dynamical complexity and fractal characteristics of Bitcoin/US Dollar and Euro/US Dollar exchange rates around the COVID-19 outbreak." *Entropy* 25, no. 2 (2023): 214.
35. Palazzi, Rafael Baptista, Gerson de Souza Raimundo Júnior, and Marcelo Cabus Klotzle. "The dynamic relationship between bitcoin and the foreign exchange market: A nonlinear approach to test causality between bitcoin and currencies." *Finance Research Letters* 42 (2021): 101893.
36. Oad Rajput, Suresh Kumar, Ishfaque Ahmed Soomro, and Najma Ali Soomro. "Bitcoin sentiment index, bitcoin performance and US dollar exchange rate." *Journal of Behavioral Finance* 23, no. 2 (2022): 150–165.
37. Erdas, Mehmet Levent, and Abdullah Emre Caglar. "Analysis of the relationships between Bitcoin and exchange rate, commodities and global indexes by asymmetric causality test." *Eastern Journal of European Studies* 9, no. 2 (2018): 27–45.
38. Szetela, Beata, Grzegorz Mentel, and Stanisław Gędek. "Dynamic econometric models." (2016) Vol. 16 (2016), 133–144. http://dx.doi.org/10.12775/DEM.2016.009.
39. Pakėnaitė, Simona, and Kamilė Taujanskaitė. "Analysis of relationship between Bitcoin emission and exchange rates of selected fiat currencies." In *International Scientific Conference "Contemporary Issues in Business, Management and Economics Engineering"*, 2019. https://doi.org/10.3846/CIBMEE.
40. Khaled, L. "The relationship between crypto currencies and official arabian currencies exchange rate." *Academy of Accounting and Financial Studies Journal* 24, no. 2 (2020): 1–10.
41. Chang, Chien-Yun, Chien-Chien Lo, Jui-Chang Cheng, Tzer-Long Chen, Liang-Yun Chi, and Chih-Cheng Chen. "Relationship between Bitcoin exchange rate and other financial indexes in time series." *Mathematical Problems in Engineering* 2021, no. 1 (2021): 8842877.
42. Baur, Dirk G., and Thomas Dimpfl. "The volatility of Bitcoin and its role as a medium of exchange and a store of value." *Empirical Economics* 61, no. 5 (2021): 2663–2683.
43. Almansour, Bashar Yaser, and M. Inairat. "The impact of exchange rates on bitcoin returns: Further evidence from a time series framework." *International Journal of Scientific & Technology Research* 9, no. 02 (2020): 4577–4581.

# *Section 2*

## *Disruptive Technologies in Business*

# 9 Intertwining between Digitalization and Triple Bottom Line Sustainability in the Banking Sector

## *Sustainable Banking Digital Ecosystem and Key Dimensions*

*Megha Garg and Parveen Kumar*

## 9.1 INTRODUCTION

The world is transitioning into a phase of digitalization during which our core efforts rely heavily on cutting-edge digital technologies, and these contemporary tools are employed in socioeconomic, environmental, sustainability, and climatic research to raise system output and effectiveness [1–4]. The advent of the digital age gave rise to promising concepts involving "the paperless office" and "telecommuting" and rapidly developing applications that were hardly foreseeable at the time. These notions have been vividly realized in the form of digital technologies like chatbots, blockchain, cloud computing, and artificial intelligence that are altering how people act, interact, and make decisions [5,6,1,7,8]. On the other hand, sustainability encompasses a broader framework [9,10] and a long-term perspective that touches triple bottom lines, namely social equality, environmental conservation, and economic prosperity [11], and makes a substantial contribution to the ideation and implementation of these three underlying pillars [12,13]. Accordingly, digitalization and sustainability are the two most influential megatrends transforming society and the economy [14–19], particularly after COVID-19 [20,8,21]. The proposition is that an enterprise that aligns with sustainability objectives like the circular economy can benefit the environment and the bottom line [22–24].

DOI: 10.1201/9781032700953-11

In this regard, Benoît Coeuré's speech at the global banking conference discussed various opportunities in financial services following the post-pandemic era. It underlined how these two hyped dimensions "green" and "digital" are interdependent rather than independent [25] as both complement one another [25,4]. Additionally, the speech asserted that central banks should make room to modify their operational frameworks to take climate-related risks within the context of their mandates. Currently, policymakers are becoming more vocal about the necessity for the banking system to design its products and services with customers while also acting pro-actively to address future social, environmental, and economic developments [26–29] to strike the ideal balance between digital finance and green innovation. Typically, technological advances in the banking sector have incentivized financial systems with an intrinsic motivation to optimize their stability and responsiveness to outside influences [30]. In this regard, Mathur and Berwa [31] unveiled the association between technology, innovation, and sustainability and tried to relate highly innovative nations and corporate organizations to perform well on sustainability-related metrics. Furthermore, banks need to orient themselves towards the outside-in and inside-out approach to embark on a sustainable transformation path to be more effective. An outside-in approach includes appealing to eco-aware clients and investors by ensuring adherence to sustainability standards. In contrast, an inside-out approach ensures that internal process optimization is compatible with sustainability.

Despite the burgeoning scholarly interest in the potential benefits and drawbacks of digitalization for sustainability [32,33,24,16,34], a comprehensive grasp of how these megatrends relate [35–37,34] in the banking sector is still absent. Even though banks have the highest percentage of IT investments ranging from 4.7% to 9.4% on average across all industries [38]. The intersection of both these disciplines indicates tremendous yet unexplored prospects to promote a shift toward sustainable development [39,16,4]. As we can witness, the Scopus database has only 21 entries with title words related to transition, sustainability, and digitalization [40]. Most importantly, previous studies were limited to specific nations or regions and lacked generalizability. Also, earlier research concentrated on the impact of digitalization on a particular dimension, i.e., the economic, social, or environmental. For instance, authors [41] assessed how bank's digitalization tactics affect financial performance in African nations. Furthermore, the authors [42] addressed the issue of fusing digital technologies and economic sustainability. They emphasized how the manufacturing cycle may made sustainable and feasible by improving connections between products and factories. In light of this, the following significant research questions need further examination:

RQ1. To explore the connection between banking digitalization and three dimensions of sustainability from an integrated perspective.

RQ2. To design a sustainable digital ecosystem for the banking sector.

RQ3. To describe the key indicators for measuring the sustainability impact of digitalization in the banking sector.

### 9.1.1 Contribution of the Study

Banks' contributions to reduce climate-related risks or to conserve resources have plenty of justifications. First, the banking industry must assume responsibility for its actions that impact three spheres, and owe this obligation to be represented in all its actions. Second, the bank's new perspective as a sustainable banker fulfils the expectations of all stakeholders [43]. Third, banks can affect the velocity and trajectory of economic growth thus playing an indispensable role in a nation's sustainable development [26,44,45]. Fourth, banks may indulge in severe disputes and scandals if the funding is utilized for polluting projects. In this respect, the viability of banks is heavily reliant on the sustainability of the businesses they lend to. Consequently, some banks consider ecological considerations when granting loans to mitigate risks [22,46]. Nevertheless, most environmental risks in the banking industry are not direct but their customers' maneuvers have a significant external influence [47,48]. Lastly, banks unambiguously have one of the largest networks, enabling them to influence consumers' actions by stepping up and rerouting capital towards investments in low-carbon and climate-resilient projects.

### 9.1.2 Paper Division

The framework of this study is as follows in light of the premises mentioned above: Section 1.1 provides the rationale for the bank's contributions to reduce climate-related risks or to conserve resources and highlights the research gap of our study. Section 2.1 succinctly explains the relationship between banking digitalization and three dimensions of sustainability. Section 3.1 represents the sustainable banking digital ecosystem, which is the main focus of this study and is reserved for describing the significant metrics for measuring sustainability impact. Finally, Section 4.1 discusses the findings and concludes the chapter with practical consequences, recommendations, limitations, and future research avenues.

## 9.2 REVIEW OF LITERATURE

### 9.2.1 Literature Showing the Interconnection between Banking Digitalization and the Economic Dimension of Sustainability

Economic indicators assess the bank's profitability, long-term sustainability, and operational effectiveness. Integrating automation tools, handling and analyzing data systems, and improving user experience technologies can all be part of the banking digital transformation. These technologies enable swift and effortless access to financial operations around the clock, enabling banks to perform better. The old business models of banks are in danger of growing digital advances in financial services since they intend to enhance user experience and produce new revenue stream [49,50]. Empirical research by Zuo et al. [51] purports that banks with more robust performance always devote additional funds to digitalizing their products, and total productivity increases by 10.7% due to purely technical factors. Recent research by Abuhasan and Moreb [52] argued that customer satisfaction and fidelity to the bank are critical drivers of ongoing banking operations, which assists in luring in more clients who are looking to acquire quality service, all of which significantly aid in boosting interest rates and bank profits.

Moreover, these technologies have strengthened banks' ability to reach all of their target markets without having to open multiple, exorbitant branches and able to cater to a broad spectrum of the customer base [53] that spans vast geographic areas, ensuing increased competitiveness and the bank's market share. According to Lugovsky [54], banks' digital strategy endeavors led to the eradication of paperwork, minimal time executing transactions and settlements, openness, and reduced operational expenditure and risk in core areas. Furthermore, Forcadell et al. [55] conclude that "corporate sustainability" could significantly strengthen banks' capacity to gain considerably from the opportunities presented by the digital revolution. The performance of banks can benefit over the long run from innovation in digital finance [56] as banks may strengthen their financial performance through digital banking [57]. Nevertheless, Scott et al. [58] analyzed the effect that SWIFT ("network-based technological infrastructure") implementation has had on bank performance. They analyzed 6,848 banks in 29 European countries and discovered that the deployment of SWIFT significantly impacts long-term profitability.

### 9.2.2 Literature Showing the Interconnection between Banking Digitalization and the Environmental Dimension of Sustainability

Environment metrics measure the influence of a bank's activities on the environment, such as carbon footprint, energy consumption, and waste production. The effective application of digital technologies can contribute to accomplishing sustainable development goals in large fractions, while future economic and environmental concerns can't be tackled sustainably without changing how existing businesses operate [59,60]. However, relying on judicious utilization of technology can optimize the material and human resources, thereby making the banks more resource-efficient [26]. In this regard, Bican and Brem [59] illuminate how environmental conservation platforms and funding for sustainable agricultural processes could help reduce agricultural carbon emissions through digital financing. Liu et al. [61] demonstrate that digital finance stimulates green innovation by eradicating financial constraints and elevating R&D spending, and in economically depressed areas and polluting industries, the effect is more acute.

Moreover, Stauropoulou and Sardianou [28] asserted that the interplay between the banking industry and the environment is leading to the generation of revolutionary financial products. Accordingly, higher sustainability standards are an outcome of digitalization, demonstrating the two megatrends' beneficial interdependencies [14,62]. In this regard, banks can create a concise and meticulous sustainability strategy that will determine their objectives and commitments and can communicate their strategy by using a variety of platforms, including websites, publications, and sustainability reports, to inform stakeholders for better reliability and transparency.

The digital revolution presents both opportunities and challenges. While it offers numerous advantages, it also brings added complexity, risks, and vulnerabilities to businesses and society [63,7]. Digital tools are significant energy consumers for banks, and the disposal of e-waste is a relevant issue. While digital technology has the potential to support green development, these two activities are not always compatible [64]. Although the simultaneous increase in the use of digital solutions may be deleterious to the environment, legislating sustainable digitalization to minimize any potential negative consequences is a feasible solution [65,25].

### 9.2.3 Literature Showing the Interconnection between Banking Digitalization and the Social Dimension of Sustainability

Social indicators evaluate the influence of banking operations on social issues, such as employee rights, diversity and equity, community involvement, and the community's well-being and healthy lifestyles. Digital finance has the potential to give underprivileged people in emerging nations the access to safe, accessible, and affordable banking services that can promote wider financial inclusion according to Ozili [56]. Furthermore, users can now adapt to digital innovations and execute the majority of tasks themselves without having to interact with bank's human resources through mobile banking or other digital applications, safeguarding both the customer and employee time as well as efforts [53,66] thereby enhancing the wellbeing of communities. In addition, it also aids in the eradication of corruption since transaction records are kept electronically, allowing for better regulation of unlawful activities [67,11] and diminishing income inequalities.

On the contrary, the widespread adoption of digital technology has expanded the scope and frequency of cyberattacks, representing a danger to the confidentiality of client data and ruining banks' reputations [68]. Additionally, providing digital financial services may be more expensive than securing clients' information across digital finance channels, and frequent data security breaches might impact customers' beliefs. However, 22% of banks worldwide have dedicated more than 25% of their annual budget to digitalized risk management, as data digitalization contributes to about 70% of digital risk [69]. The fundamental issue in regulating cyberattack threats for banking organizations is specific legal and supervisory requirements [70]. Henceforth, the most significant bottleneck is to safeguard investor and customer data privacy along with financial stability from a compliance perspective.

## 9.3 RESEARCH METHODOLOGY

This review article applied a qualitative approach. A literature review method appears to be a promising strategy since it is an essential stage in defining a study area and aids in locating the field's conceptual substance. The purpose of the review was to collect literature on the interconnection between two megatrends and to get an insight into how sustainability emerges from banking digitalization from the body of existing literature. To accomplish this, an exhaustive search was carried out from reputed databases like "Scopus" and "Web of Science" [71] to determine the parameters of the study and give a broader perspective of the phenomenon. A meticulous examination of the literature will assist us in designing a sustainable digital ecosystem and help us pinpoint vital indicators, potential issues, and areas that still require investigation.

### 9.3.1 Sustainable Banking Digital Ecosystem

Banks' conventional business structures, procedures, and products are now being fundamentally reoriented due to the growing grip of technological influence and the emergence of new banking providers (FinTechs), increasing the likelihood that banks will face disruption if they do not change. The entire economic system and financial services will be in jeopardy if such changes are overlooked by financial institutions like banks. Consequently, the banking industry must adjust to these market shifts as an unavoidable requirement [72]. FinTech businesses worldwide are revolutionizing how people manage their money, further disrupting the market. Live chatbots are fully operational, banking apps are altering mobile payment systems, and interfaces are redefining conventional banking structures. Henceforth, the cost of financial intermediation has decreased subsequently under banking digital transformation that has also opened up a digital ecosystem that enables bank staff to deliver better customer support and entice new clients.

The digital banking ecosystem has five key enablers: customer centricity, execution capabilities, organizational foundation, talent pooling, and collaboration. Banks are now redesigning their organizations and employee positions to streamline multiple digital banking undertakings, as they need to do so to avoid losing their training investments. Banking professionals are gradually expected to be tech-savvy with the following skill sets: agile thinking, risk & governance, human-centered design, data-driven decision-making, and digital awareness. Generally, a majority bank wants to develop organizational agility since it promotes innovation and talent upskilling to meet ecosystem needs. To succeed over the long haul, embracing enduring human skills is critical. When banks involve their customers in the design process to meet their specifications, the resulting end-product holds more value to customers. In addition, banks may target their communications to the correct audience by using a data-driven approach to segment users. Nonetheless, increased interconnection and new technologies within the banking ecosystem pose significant cyber risks and imperil the bank's control over the client experience. Therefore, having the necessary skills to comprehend and handle such risks during the digitalization process is crucial. Comprehensive utilization of a broad spectrum of digital technologies, including artificial technologies, blockchain, augmented reality, chatbots, cloud banking, biometrics, big data, machine learning, and others, in the banking business enables swift and flexible infrastructure, diminution of risk, simplified payment procedure, and instantaneous lending [73,11].

The sustainable banking digital ecosystem (Figure 9.1) designed with the above-mentioned digitalization enablers, future-enabled skill set, and technologies may monitor, organize, and synchronize the banking activities: reduction in operational cost and risk, accessibility, and uninterruptible service; user retention and satisfaction through digital banking examine real-time patterns; evaluate the massive amount of data; fraud anticipation through artificial intelligence; defining enormous amounts of structured and unstructured data to predict consumer behavior through big data, voice interface, user feedback, solving queries through chatbots, and more.

The pluses produced through digitalizing the entire banking system process can be distilled into three main categories: operational efficiency, resource efficiency, and community well-being. However, a sustainable digital banking ecosystem is shown in Figure 9.1, which exemplifies how digitalization affects banking from the social, environmental, and economic facets of sustainability. Below are the three trends that are prompting the banking sector to go for sustainability:

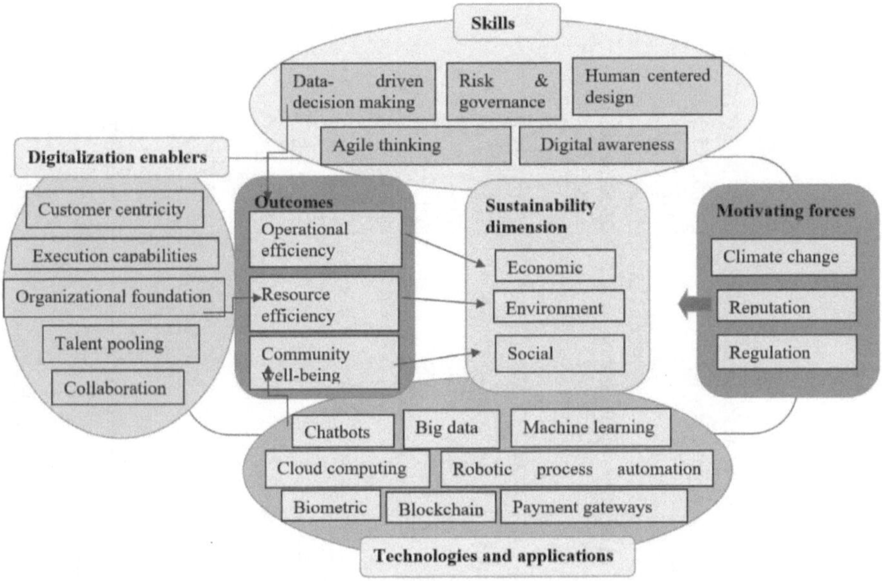


**FIGURE 9.1** Sustainable banking digital ecosystem (author's compilation).

i. Climate change: Due to the constant risk, the frequent repercussions of climate change and sustainability have become critical for banks. Consequently, shareholders, bank employees, and consumers urge banks to adopt environmental pledges to lessen the effects of climate change.
ii. Reputation: Banks are under obligation to operate sustainably and provide full disclosure about how it is done as a result of rapidly evolving user awareness and preferences. Banks that set weak sustainability goals run the risk of receiving negative public attention.
iii. Regulation: The financial sector is slammed with new laws and reporting requirements related to climate risk. Therefore, financial institutions must prioritize their environmental goals and change their business practices quickly to match a more climate-friendly narrative.

### 9.3.2 Key Indicators for Measuring Sustainability Impact

Table 9.1 discusses different metrics used to assess the economic, environmental, and social sustainability resulting from the digitalization of banking. After thorough consideration, 24 relevant criteria have been identified. Of these, 11 pertain to the economic aspect of sustainability, 6 to the environmental aspect, and 7 to the social aspect.

## 9.4 CONCLUSION

The evolving significance of sustainable development and the transformation towards a lower-carbon economy will necessitate an upswing of ingenious digital resolutions

**TABLE 9.1**
**Key indicators for measuring the sustainability impact**

| Sustainability dimensions | A. Indicators | B. Description | C. Source |
|---|---|---|---|
| D. Economic | Performance | Change in financial performance of banks in terms of profitability, liquidity, and financial stability | [74,58,57,56,51,73,50,41] |
| | Productivity | Change in bank productivity in terms of work, production, and management efficiency | [75,66,76,51] |
| | Economic stability | Change in the gross domestic product of the digitalized economy | [56,77,78,4] |
| | Revenue streams | Opening up new revenue opportunities | [49,50] |
| | Operational efficiency | Change in operational efficiencies in terms of automation, data management, and integration | [26,51,54,66,79] |
| | Accuracy | Change in terms of financial or human errors | [73,80] |
| | Customer base | Change in existing customer base | [53] |
| | Competitive edge | Change in competitive confrontation in terms of cost reduction and technology adoption | [7,81,51,73,76] |
| | Market share | Change in bank market share in local, regional, and global market | [81,51] |
| | Risk | Change in operational risk | [79,54] |
| | Interest rate | Change in market interest rate | [52] |
| Environment | Green innovation | Impact of digital financing on green innovation | [82,61] |
| | Carbon footprint | Change in the amount of carbon emission | [59,83] |
| | Pollution | Change in pollution emissions by investing in climate-resilient projects | [22,46] |
| | Sustainability | Integration of technology and procedures in sustainable development | [14,28,82,62,4,84] |
| | Resource efficiency | Change in the amount of paper waste in terms of saving the environment | [26,66,79,4] |
| | E-waste | Change in hazardous waste disposal | [85,64] |
| Social | Financial inclusion | Change in terms of economic development, employment rate, and financial access | [86,56,77] |
| | Well-being | Change in terms of time as well as efforts to execute tasks | [53,66,79] |
| | Experience | Change in terms of customer satisfaction and loyalty | [49,50] |
| | Cyber risk | Change in cyber-crime activities in terms of data intrusion, confidentiality, or identity theft | [70,87,68,79] |
| | Corruption | Regulation of unlawful activities and improvement in tax collection | [67] |
| | Impact | The social impact generated through banking digitalization | [72] |
| | Acceptance | Traditional business models, financial institutions, and community acceptance of digital applications | [72,79] |

that support green endeavors and simplify the transition towards a sustainable financial system. Henceforth, our article has attempted to highlight the pivotal connection between the two gigantic trends, digitalization and sustainability, and addressed the following questions: What are the characteristics of digitalization and related technologies in the banking network? How has the implementation of digitalization altered banking procedures? And what are the metrics for measuring the sustainability impact of banking digitalization? Banking digitalization has yet to achieve a mature state, given that it is still in its nascent stages of development. Because of this, sustainability implications can still be modified and optimized over time with the maturity level of digitalization. The exhaustive literature analysis makes it abundantly evident that the fate of one is dependent on another and affirms that banking digitalization has enormous potential to contribute to the accomplishment of the sustainability concept, which can assist decision-makers in better comprehending emerging trends in the implementation of digitalization and the degree to which these tendencies facilitate the shift to greater sustainability [84].

The findings of the study confirmed that exploiting digital technologies and applications in the banking segment substantially impacted sustainability. Predominantly, the sustainability impact of digitalization concerning the economic dimension was more important than the other dimensions (Figure 9.1). However, the financial implication can be seen in terms of performance, economic stability, accuracy, productivity, and market share, and has a positive impact in general. Besides this, environmental implication significantly reduces stationary cost, carbon, and greenhouse gas emissions, but at the same time, disposal of e-waste is a separate alarming issue. The social impact of digitalization had a negative effect in terms of customer privacy, confidentiality, and cyber risk. Nevertheless, financial education and awareness can be improved through digitalization in banking.

However, it is relevant to emphasize that in a country with such a diverse population as India, it is imperative to look into the security issues of digitalization. It is an ongoing process that requires constant watchfulness for technical mistakes, data security problems, and cyber disruptions [87]. Accordingly, it is recommended that financial institutions cultivate new partnerships with stakeholders to mitigate the risks associated with cyberattacks and attract more investments to manage cyber threats to support the financial sector's sustainable growth. In this regard, government and research practitioners are accountable for building new digital tools that could support sustainability or manage digital risk. Moreover, the government must enact national legislation on handling e-waste and encourage collaboration between local governments, companies/producers, and recycling industries to establish a Web based e-waste collecting system [88].

### 9.4.1 Implications

The study suggests the following relevant policy implications in light of the above-mentioned findings. First, the consequences of this research are meaningful for regulators and officials, who must deal with sustainability and digital transformation while endeavoring to take advantage of any synergies between the two phenomena in the banking domain. Second, this research answers a practical call to comprehend how to employ banking digital technologies to accomplish sustainability-related metrics. Third, various indicators that have been identified for measuring three dimensions of sustainability (Table 9.1) would help the research scholars and practitioners from the management field to measure the impact holistically. Fourth, this research act as an organizing framework for further empirical investigations that leverage the metrics stumbled upon (Table 9.1). Fifth, this interconnection will not only broaden practitioners' knowledge on the subject but also educate and deliver them with access to the particulars of the current academic production, which has been characterized and analyzed for the first time in this article. Lastly, the study holds profound consequences for policymakers and practitioners, especially towards envisaged positives that digital technologies could bring to the banking industry's efforts to meet the 2030 sustainable development goals.

### 9.4.2 Limitations of the Study

Despite its significant ramifications, some limits should be emphasized as they may present valuable avenues for further research. The first restriction is that sustainability and digitalization are multidimensional, extremely complicated, and constantly growing fields that need strong theory development, broad definitions, and unambiguous research boundary delineation. As noted, the literature paints a picture of a young, disjointed world of unresolved conflicts. Second, the main shortcoming is the lack of a quantitative component, as it has only qualitatively evaluated the economic, environmental, and social dimensions of sustainability. Culture can also be the fourth pillar of sustainability [89] along with technical, political, and legal dimensions [90]. Furthermore, we acknowledge that some pertinent research may have been unintentionally left out due to the evolving landscape of the subject matter and the methodology that relies on preliminary screening and qualitative assessments.

### 9.4.3 Future Research Directions

To add value to our investigation, specific attention can be given to the practices and strategies addressing sustainability in the context of banking digital innovation. Moreover, digitalization for sustainability can be explored alongside other prominent financial themes such as sustainable reporting, risk/sustainability trade-offs, and digital innovation or value creation. However, to promote a sustainable and informed business environment, the principal objective of sustainability reporting is to safeguard and stimulate public trust in corporations (Diener & Špaček, 2020). Moreover, more inductive research techniques may also offer supplementary insights. For example, "Which sustainable development goals can be purely achieved through banking digital transformation?" would clarify our study more clearly. In addition, the future outlook can consider other sustainability dimensions such as culture, governance, technical, legal, and political to measure overall indicators for evaluating the sustainability impact of digitalization in banking. Importantly, the nexus could be explored in detail by applying a range of parametric and non-parametric techniques, including linear and quantile regression. Lastly, a cohesive viewpoint and overarching theoretical frameworks are required for future study advancements. The identified gaps in knowledge are anticipated to encourage study with more targeted outcomes from management scholars to the practical community.

## REFERENCES


1. Colbert, Amy, Nick Yee, and Gerard George. "The digital workforce and the workplace of the future." *Academy of Management Journal* 59, no. 3 (2016): 731–739. https://doi.org/10.5465/AMJ.2016.4003
2. Balogun, Abdul-Lateef, Danny Marks, Richa Sharma, Himanshu Shekhar, Chiden Balmes, Dikman Maheng, Adnan Arshad, and Pourya Salehi. "Assessing the potentials of digitalization as a tool for climate change adaptation and sustainable development in urban centres." *Sustainable Cities and Society* 53 (2020): 101888. https://doi.org/10.1016/j.scs.2019.101888
3. Ceipek, René, Julia Hautz, Antonio Messeni Petruzzelli, Alfredo De Massis, and Kurt Matzler. "A motivation and ability perspective on engagement in emerging digital technologies: The case of internet of things solutions." *Long Range Planning* 54, no. 5 (2021): 101991. https://doi.org/10.1016/j.lrp.2020.101991
4. Xu, Jiaqi, Shengxiang She, and Wen Liu. "Role of digitalization in environment, social and governance, and sustainability: Review-based study for implications." *Frontiers in Psychology* 13 (2022): 961057. https://doi.org/10.3389/fpsyg.2022.961057
5. Hammi, Mohamed Tahar, Badis Hammi, Patrick Bellot, and Ahmed Serhrouchni. "Bubbles of trust: A decentralized blockchain-based authentication system for IoT." *Computers & Security* 78 (2018): 126–142. https://doi.org/10.1016/J.COSE.2018.06.004
6. Jarrahi, Mohammad Hossein. "Artificial intelligence and the future of work: Human-AI symbiosis in organizational decision making." *Business Horizons* 61, no. 4 (2018): 577–586. https://doi.org/10.1016/J.BUSHOR.2018.03.007
7. Brenner, Barbara. "Transformative sustainable business models in the light of the digital imperative—A global business economics perspective." *Sustainability* 10, no. 12 (2018): 4428. https://doi.org/10.3390/SU10124428
8. Faraj, Samer, Wadih Renno, and Anand Bhardwaj. "Unto the breach: What the COVID-19 pandemic exposes about digitalization." *Information and Organization* 31, no. 1 (2021): 100337. https://doi.org/10.1016/J.INFOANDORG.2021.100337
9. Beier, Grischa, Silke Niehoff, and Bing Xue. "More sustainability in industry through industrial internet of things?" *Applied Sciences* 8, no. 2 (2018): 219. https://doi.org/10.3390/APP8020219
10. Ghobakhloo, Morteza. "Industry 4.0, digitization, and opportunities for sustainability." *Journal of Cleaner Production* 252 (2020): 119869. https://doi.org/10.1016/j.jclepro.2019.119869
11. Garg, Megha, and Parveen Kumar. "Harnessing digital technologies for triple bottom line sustainability in the banking industry: A bibliometric review." *Future Business Journal* 10, no. 1 (2024): 62. https://doi.org/10.1186/s43093-024-00336-2
12. Choi, Sungchul, and Alex Ng. "Environmental and economic dimensions of sustainability and price effects on consumer responses." *Journal of Business Ethics* 104 (2011): 269–282. https://doi.org/10.1007/s10551-011-0908-8
13. Ford, Simon, and Mélanie Despeisse. "Additive manufacturing and sustainability: an exploratory study of the advantages and challenges." *Journal of Cleaner Production* 137 (2016): 1573–1587. https://doi.org/10.1016/J.JCLEPRO.2016.04.150
14. Aksin-Sivrikaya, Sezen, and C. B. Bhattacharya. "Where digitalization meets sustainability: Opportunities and challenges." In T. Osburg & C. Lohrmann (Eds.), *Sustainability in a Digital World. CSR, Sustainability, Ethics & Governance*. Cham: Springer. https://doi.org/10.1007/978-3-319-54603-2_3.

15. Kiron, David, and Gregory Unruh. "The convergence of digitalization and sustainability." *MIT Sloan Management Review*, 2018. https://sloanreview.mit.edu/article/the-convergence-of-digitalization-and-sustainability/.
16. Del Río Castro, Gema, María Camino González Fernández, and Ángel Uruburu Colsa. "Unleashing the convergence amid digitalization and sustainability towards pursuing the Sustainable Development Goals (SDGs): A holistic review." *Journal of Cleaner Production* 280 (2021): 122204. https://doi.org/10.1016/j.jclepro.2020.122204
17. Dell'Erba, Marco. "Sustainable digital finance and the pursuit of environmental sustainability." In D. Busch, G. Ferrarini, & S. Grünewald (Eds.), *Sustainable Finance in Europe. EBI Studies in Banking and Capital Markets Law*. Cham: Palgrave Macmillan. https://doi.org/10.1007/978-3-031-53696-0_3
18. Lichtenthaler, Ulrich. "Digitainability: The combined effects of the megatrends digitalization and sustainability." *Journal of Innovation Management* 9, no. 2 (2021): 64–80. https://doi.org/10.24840/2183-0606_009.002_0006
19. Guandalini, Ilaria. "Sustainability through digital transformation: A systematic literature review for research guidance." *Journal of Business Research* 148 (2022): 456–471. https://doi.org/10.1016/j.jbusres.2022.05.003
20. Almeida, Fernando, José Duarte Santos, and José Augusto Monteiro. "The challenges and opportunities in the digitalization of companies in a post-COVID-19 World." *IEEE Engineering Management Review* 48, no. 3 (2020): 97–103. https://doi.org/10.1109/EMR.2020.3013206
21. Orzeł, Bartosz, and Radosław Wolniak. "Digitization in the design and construction industry—remote work in the context of sustainability: A study from Poland." *Sustainability* 14, no. 3 (2022): 1332. https://doi.org/10.3390/su14031332
22. Weber, Olaf, Roland W. Scholz, and Georg Michalik. "Incorporating sustainability criteria into credit risk management." *Business Strategy and the Environment* 19, no. 1 (2010): 39–50. https://doi.org/10.1002/BSE.636
23. Bocken, Nancy M. P., Samuel W. Short, Padmakshi Rana, and Steve Evans. "A literature and practice review to develop sustainable business model archetypes." *Journal of Cleaner Production* 65 (2014): 42–56. https://doi.org/10.1016/J.JCLEPRO.2013.11.039
24. Geissdoerfer, Martin, Paulo Savaget, Nancy MP Bocken, and Erik Jan Hultink. "The circular economy–A new sustainability paradigm?" *Journal of Cleaner Production* 143 (2017): 757–768. https://doi.org/10.1016/J.JCLEPRO.2016.12.048
25. Carstens, Agustín. "Central bank digital currencies: putting a big idea into practice." *Speech, BIS Innovation Summit*, Bank for International Settlements, June 22, 2021. https://www.bis.org/speeches/sp210622a.htm
26. Ramnarain, Taruna Devi, and Mahdevi Tiagarassa Pillay. "Designing sustainable banking services: The case of Mauritian banks." *Procedia-Social and Behavioral Sciences* 224 (2016): 483–490. https://doi.org/10.1016/J.SBSPRO.2016.05.424
27. Azevedo, Susana Garrido, Radu Godina, and João Carlos de Oliveira Matias. "Proposal of a sustainable circular index for manufacturing companies." *Resources* 6, no. 4 (2017): 63. https://doi.org/10.3390/RESOURCES6040063
28. Stauropoulou, Athanasia, and Eleni Sardianou. "Understanding and measuring sustainability performance in the banking sector." In *IOP Conference Series: Earth and Environmental Science*, vol. 362, no. 1, p. 012128. IOP Publishing, 2019. https://doi.org/10.1088/1755-1315/362/1/012128
29. Arner, Douglas W., Ross P. Buckley, Dirk A. Zetzsche, and Robin Veidt. "Sustainability, FinTech and financial inclusion." *European Business Organization Law Review* 21 (2020): 7–35. https://doi.org/10.1007/s40804-020-00183-y
30. Tsindeliani, Imeda A., Maxim M. Proshunin, Tatyana D. Sadovskaya, Zhanna G. Popkova, Mariam A. Davydova, and Oksana A. Babayan. "Digital transformation of the banking system in the context of sustainable development." *Journal of Money*

*Laundering Control* 25, no. 1 (2022): 165–180. https://doi.org/10.1108/JMLC-02-2021-0011

31. Mathur, Kalpana, and Akanksha Berwa. "Sustainable competitiveness: Redefining the future with technology and innovation." *Journal of Sustainable Finance & Investment* 7, no. 3 (2017): 290–306. https://doi.org/10.1080/20430795.2017.1300855
32. Seele, Peter. "Envisioning the digital sustainability panopticon: a thought experiment of how big data may help advancing sustainability in the digital age." *Sustainability Science* 11 (2016): 845–854. https://doi.org/10.1007/s11625-016-0381-5
33. Seele, Peter, and Irina Lock. "The game-changing potential of digitalization for sustainability: possibilities, perils, and pathways." *Sustainability Science* 12 (2017): 183–185. https://doi.org/10.1007/s11625-017-0426-4
34. Brenner, Barbara, and Barbara Hartl. "The perceived relationship between digitalization and ecological, economic, and social sustainability." *Journal of Cleaner Production* 315 (2021): 128128. https://doi.org/10.1016/j.jclepro.2021.128128
35. Boone, Christopher A., Joseph B. Skipper, and Benjamin T. Hazen. "A framework for investigating the role of big data in service parts management." *Journal of Cleaner Production* 153 (2017): 687–691. https://doi.org/10.1016/J.JCLEPRO.2016.09.201
36. Song, Malin, Ling Cen, Zhixia Zheng, Ron Fisher, Xi Liang, Yutao Wang, and Donald Huisingh. "How would big data support societal development and environmental sustainability? Insights and practices." *Journal of Cleaner Production* 142 (2017): 489–500. https://doi.org/10.1016/J.JCLEPRO.2016.10.091
37. Hommel, Kristin, and Peter M. Bican. "Digital entrepreneurship in finance: Fintechs and funding decision criteria." *Sustainability* 12, no. 19 (2020): 8035. https://doi.org/10.3390/SU12198035
38. Gopalan, Sai, Gaurav Jain, Gaurav Kalani, and Jessica Tan. "Breakthrough IT banking." *McKinsey Q* 26 (2012): 30–35.
39. Osburg, Thomas. "Sustainability in a digital world needs trust." In *Sustainability in a Digital World*, edited by Thomas Osburg and Christian Lohrmann, 1–14. CSR, Sustainability, Ethics & Governance. Cham: Springer, 2017. https://doi.org/10.1007/978-3-319-54603-2_1
40. Andersen, Allan Dahl, Koen Frenken, Victor Galaz, Florian Kern, Laurens Klerkx, Matthijs Mouthaan, Laura Piscicelli, Juliet B. Schor, and Taneli Vaskelainen. "On digitalization and sustainability transitions." *Environmental Innovation and Societal Transitions* 41 (2021): 96–98. https://doi.org/10.1016/j.eist.2021.09.013
41. Theiri, Saliha, and Slim Hadoussa. "Digitization effects on banks' financial performance: The case of an African country." *Competitiveness Review: An International Business Journal* ahead-of-print (2023). https://doi.org/10.1108/CR-10-2022-0147
42. Cricelli, Livio, and Serena Strazzullo. "The economic aspect of digital sustainability: A systematic review." *Sustainability* 13, no. 15 (2021): 8241. https://doi.org/10.3390/su13158241
43. Rebai, Sonia, Mohamed Naceur Azaiez, and Dhafer Saidane. "Sustainable performance evaluation of banks using a multi-attribute utility model: An application to French banks." *Procedia Economics and Finance* 2 (2012): 363–372.
44. Zhixia, Chen, Md Miraj Hossen, Sayed Sami Muzafary, and Mareum Begum. "Green banking for environmental sustainability-present status and future agenda: Experience from Bangladesh." *Asian Economic and Financial Review* 8, no. 5 (2018): 571–585. https://doi.org/10.18488/journal.aefr.2018.85.571.585
45. Sharma, Meenakshi, and Akanksha Choubey. "Green banking initiatives: a qualitative study on Indian banking sector." *Environment, Development and Sustainability* 24, no. 1 (2022): 293–319. https://doi.org/10.1007/s10668-021-01426-9
46. Moufty, Souad, Ephraim Clark, and Basil Al-Najjar. "The different dimensions of sustainability and bank performance: Evidence from the EU and the USA." *Journal*

*of International Accounting, Auditing and Taxation* 43 (2021): 100381. https://doi.org/10.1016/j.intaccaudtax.2021.100381
47. Nath, Vikas, Nitin Nayak, and Ankit Goel. "Green banking practices–A review." *IMPACT: International Journal of Research in Business Management (IMPACT: IJRBM)* 2 (2014): 45–62.
48. Garg, Megha, and Parveen Kumar. "Do Banks' employees have different attitudes in the public-vs private-sector towards green banking?" *Journal of General Management Research* 10, no. 1 (2022): 1–13.
49. Svahn, Fredrik, Lars Mathiassen, and Rikard Lindgren. "Embracing digital innovation in incumbent firms." *MIS Quarterly* 41, no. 1 (2017): 239–254. https://doi.org/10.25300/MISQ/2017/41.1.12
50. Naimi-Sadigh, Ali, Tayebeh Asgari, and Mohammad Rabiei. "Digital transformation in the value chain disruption of banking services." *Journal of the Knowledge Economy* 13, no. 2 (2022): 1212–1242. https://doi.org/10.1007/s13132-021-00759-0
51. Zuo, Lihua, Jack Strauss, and Lijuan Zuo. "The digitalization transformation of commercial banks and its impact on sustainable efficiency improvements through investment in science and technology." *Sustainability* 13, no. 19 (2021): 11028. https://doi.org/10.3390/SU131911028
52. Abuhasan, Firas, and Mohammed Moreb. "The impact of the digital transformation on customer experience in Palestine banks." In *2021 International Conference on Information Technology (ICIT)*, pp. 43–48. IEEE, 2021. https://doi.org/10.1109/ICIT52682.2021.9491744
53. Jayawardhena, Chanaka, and Paul Foley. "Changes in the banking sector–the case of Internet banking in the UK." *Internet Research* 10, no. 1 (2000): 19–31. https://doi.org/10.1108/10662240010312048
54. Lugovsky, Vadim. "Digital Transformation in Banking: How to Make the Change." *Forbes*, December 29, 2021. https://www.forbes.com/sites/forbesbusinesscouncil/2021/12/29/digital-transformation-in-banking-how-to-make-the-change/?sh=3d36fb926999
55. Forcadell, Francisco Javier, Elisa Aracil, and Fernando Úbeda. "The impact of corporate sustainability and digitalization on international banks' performance." *Global Policy* 11 (2020): 18–27. https://doi.org/10.1111/1758-5899.12761
56. Ozili, P. K. "Impact of digital finance on financial inclusion and stability." *Borsa Istanbul Review* 18, no. 4 (2018): 329–340. https://doi.org/10.1016/J.BIR.2017.12.003
57. Mbama, Cajetan Ikechukwu, Patrick Ezepue, Lyuba Alboul, and Martin Beer. "Digital banking, customer experience and financial performance: UK bank managers' perceptions." *Journal of Research in Interactive Marketing* 12, no. 4 (2018): 432–451. https://doi.org/10.1108/JRIM-01-2018-0026
58. Scott, Susan V., John Van Reenen, and Markos Zachariadis. "The long-term effect of digital innovation on bank performance: An empirical study of SWIFT adoption in financial services." *Research Policy* 46, no. 5 (2017): 984–1004. https://doi.org/10.1016/J.RESPOL.2017.03.010
59. Bican, Peter M., and Alexander Brem. "Digital business model, digital transformation, digital entrepreneurship: Is there a sustainable "digital"?" *Sustainability* 12, no. 13 (2020): 5239. https://doi.org/10.3390/SU12135239
60. Diener, Florian, and Miroslav Špaček. "Digital transformation in banking: A managerial perspective on barriers to change." *Sustainability* 13, no. 4 (2021): 2032. https://doi.org/10.3390/su13042032
61. Liu, Jiamin, Yalin Jiang, Shengdao Gan, Ling He, and Qingfeng Zhang. "Can digital finance promote corporate green innovation?" *Environmental Science and Pollution Research* 29, no. 24 (2022): 35828–35840. https://doi.org/10.1007/s11356-022-18667-4
62. Gupta, Shivam, Mahsa Motlagh, and Jakob Rhyner. "The digitalization sustainability matrix: A participatory research tool for investigating digitainability." *Sustainability* 12, no. 21 (2020): 9283. https://doi.org/10.3390/SU12219283

63. Scholz, R. W. "Digital threat and vulnerability management: The SVIDT method. *Sustainability* 9, no. 4 (2017): 554. https://doi.org/10.3390/SU9040554
64. Peng, Grace. "Digitalization in banking: How to really reap the benefits." *World Economic Forum*, May 2022. https://www.weforum.org/agenda/2022/05/heres-how-to-really-reap-the-benefits-of-the-digitalization-in-banking-trend/
65. Stuermer, Matthias, Gabriel Abu-Tayeh, and Thomas Myrach. "Digital sustainability: Basic conditions for sustainable digital artifacts and their ecosystems." *Sustainability Science* 12 (2017): 247–262. https://doi.org/10.1007/S11625-016-0412-2
66. Mathapati, Chandrashekar M. "Key aspects of digital technology and its impact on bank employee performance." *IOSR Journal of Business and Management (IOSR-JBM) e-ISSN* 23, no. 8 (2021): 01–03. https://doi.org/10.9790/487X-2308050103
67. Prakash, C., R. Yadav, A. Singh, and K. Yadav. "Modeling the entrepreneurial intention among business students." *Pacific Business Review International* 14, no. 3 (2021).
68. Prakash, C., R. Yadav, and S. Kadyan. "Effect of the price drop on customer's perceived evaluation across selected product categories." *Journal of Revenue and Pricing Management* 20 (2021): 204–210.
69. McKinsey & Company. "The future of risk management in the digital era." *McKinsey & Company*, 2017. https://www.mckinsey.com/~/media/mckinsey/business%20functions/risk/our%20insights/the%20future%20of%20risk%20management%20in%20the%20digital%20era/future-of-risk-management-in-the-digital-era-iif-and-mckinsey.pdf
70. Crisanto, Juan Carlos, and Jermy Prenio. *Regulatory Approaches to Enhance Banks' Cyber-Security Frameworks.* Bank for International Settlements, Financial Stability Institute, 2017.
71. Garg, Megha, and Parveen Kumar. "Social entrepreneurship and sustainable development: A bibliometric analysis." *Colombo Business Journal* 14, no. 2 (2023): 99–131. https://doi.org/10.4038/cbj.v14i2.159
72. Diener, Florian, and Miroslav Špaček. "The role of 'digitalization'in German sustainability bank reporting." *International Journal of Financial Studies* 8, no. 1 (2020): 16. https://doi.org/10.3390/ijfs8010016
73. Balkan, Bülent. "Impacts of digitalization on banks and banking." *The Impact of Artificial Intelligence on Governance, Economics and Finance* I (2021): 33–50. https://doi.org/10.1007/978-981-33-6811-8_3
74. Oyewole, O., M. Abba, … J. G.-I. J. of, & 2013, undefined. "E-banking and bank performance: Evidence from Nigeria." *Indianjournals.Com* 2 (2013): 766–771.
75. Casolaro, Luca, and Giorgio Gobbi. "Information technology and productivity changes in the banking industry." *Economic Notes* 36, no. 1 (2007): 43–76. https://doi.org/10.1111/J.1468-0300.2007.00178.X
76. Rashwan, Abdul Rahman Mohammed Suleiman, and Zainab Abd-Elhafiz Ahmed Kassem. "The role of digital transformation in increasing the efficiency of banks' performance to enhance competitive advantage." In A. M. A. Musleh Al-Sartawi (Eds.), *The Big Data-Driven Digital Economy: Artificial and Computational Intelligence*, 325–335. Cham: Springer International Publishing, 2021. https://doi.org/10.1007/978-3-030-73057-4_25
77. Lutfi, Abdalwali, Manaf Al-Okaily, Malek Hamed Alshirah, Ahmad Farhan Alshira'h, Thaer Ahmad Abutaber, and Manal Ali Almarashdah. "Digital financial inclusion sustainability in Jordanian context." *Sustainability* 13, no. 11 (2021): 6312. https://doi.org/10.3390/su13116312
78. Tiutiunyk, Inna, Josef Drabek, Nataliia Antoniuk, Valentinas Navickas, and Pavlo Rubanov. "The impact of digital transformation on macroeconomic stability: Evidence from EU countries." *Journal of International Studies (2071–8330)* 14, no. 3 (2021): 220–234. https://doi.org/10.14254/2071-8330.2021/14-3/14
79. Kitsios, Fotis, Ioannis Giatsidis, and Maria Kamariotou. "Digital transformation and strategy in the banking sector: Evaluating the acceptance rate of e-services." *Journal of*

*Open Innovation: Technology, Market, and Complexity* 7, no. 3 (2021): 204. https://doi.org/10.3390/JOITMC7030204
80. Remondino, Marco, and Alessandro Zanin. "Logistics and agri-food: Digitization to increase competitive advantage and sustainability. Literature review and the case of Italy." *Sustainability* 14, no. 2 (2022): 787. https://doi.org/10.3390/su14020787
81. Hadi, Anmar Muzaffar, and Salam Jassim Hmood. "Analysis of the role of digital transformation strategies in achieving the edge of financial competition." *International Journal of Innovation, Creativity and Change* 10, no. 11 (2020): 19–40.
82. El Hilali, Wail, Abdellah El Manouar, and Mohammed Abdou Janati Idrissi. "Reaching sustainability during a digital transformation: A PLS approach." *International Journal of Innovation Science* 12, no. 1 (2020): 52–79. https://doi.org/10.1108/IJIS-08-2019-0083
83. Chang, Jianxin. "The role of digital finance in reducing agricultural carbon emissions: evidence from China's provincial panel data." *Environmental Science and Pollution Research* 29, no. 58 (2022): 87730–87745. https://doi.org/10.1007/s11356-022-21780-z
84. Ionescu-Feleagă, Liliana, Bogdan-Ştefan Ionescu, and Oana Cristina Stoica. "The link between digitization and the sustainable development in European union countries." *Electronics* 12, no. 4 (2023): 961. https://doi.org/10.3390/electronics12040961
85. Murugesan, San. "Harnessing green IT: Principles and practices." *IT Professional* 10, no. 1 (2008): 24–33. https://doi.org/10.1109/MITP.2008.10
86. Allen, Franklin, Asli Demirguc-Kunt, Leora Klapper, and Maria Soledad Martinez Peria. "The foundations of financial inclusion: Understanding ownership and use of formal accounts." *Journal of Financial Intermediation* 27 (2016): 1–30. https://doi.org/10.1016/J.JFI.2015.12.003
87. Saini, C. P. "Difficulty in transition: Study on hindrances in online shopping for potential shoppers." *International Journal of Online Marketing (IJOM)* 10, no. 2 (2020): 17–29.
88. Rimantho, Dino, Syaiful Syaiful, Nurfaida, and Uci Sulandari. "Electronic waste bank model as a solution for implementing circular economy: Case study DKI Jakarta-Indonesia." *Frontiers in Built Environment* 8 (2022): 1030196. https://doi.org/10.3389/FBUIL.2022.1030196
89. Nurse, Keith. "Culture as the fourth pillar of sustainable development." *Small States: Economic Review and Basic Statistics* 11 (2006): 28–40.
90. Pawłowski, A. "How many dimensions does sustainable development have?" *Sustainable Development* 16, no. 2 (2008): 81–90. https://doi.org/10.1002/SD.339

# 10 Navigating Soft Technology Disruption

## *Transforming Accounting Education for Industry 4.0 and 5.0 Era*

*Amardeep Singh, Amit Dangi, Chand Prakash, Neha Gupta, and K Tara Shankar*

## 10.1 INTRODUCTION

Industry 4.0 and its further evolution into Industry 5.0 have brought significant changes in almost all sectors, including manufacturing, services, and other areas. Industry 4.0 is a paradigm change defined, among other advanced technologies, by the combination of cloud computing, cognitive computing, the Internet of Things (IoT), and cyber-physical systems [1]. In industrial processes, this integration promotes connectivity and real-time data sharing, opening the door to automation, predictive analytics, and wise decision-making [2]. Industry 4.0 is envisioned as a comprehensive framework for rethinking industrial operations, business models, and value chains, going beyond technology innovations [3]. Industry 4.0 aims to improve operational efficiency, agility, and sustainability while enabling innovative product and service offerings by utilizing digitalization and data-driven approaches [4].

The emergence of Industry 4.0 has led to significant transformation across multiple sectors, including accounting, calling for a re-evaluation of conventional methods and procedures. In the digital age, accounting professionals must deal with changing expectations and demands, which has resulted into reassessment of the skill sets and competencies necessary for success in the industry [5]. The convergence of Industry 4.0 and environmental accounting signals a new revolution in accounting procedures [6]. To optimize the advantages of sustainable development, professional associations and governmental organizations are urged to assist in the integration of Industry 4.0 technology into environmental accounting practice and education.

It's critical to accept technology innovations while upholding the core accounting principles as the accounting profession navigates the effects of Industry 4.0 and 5.0. Professionals are attempting to effectively navigate the digital landscape and produce sustainable company outcomes by using the promise of Industry 4.0

DOI: 10.1201/9781032700953-12

and 5.0 technologies and modifying accounting processes to meet growing industry demands. Therefore, accounting procedures are about to change dramatically in the context of Industry 4.0 and 5.0, requiring a paradigm change in the way financial data is handled, examined, and presented. The businesses may be able to make better decisions and operate more efficiently and transparently by incorporating advanced technology like blockchain, artificial intelligence (AI), and big data analytics into accounting procedures.

Accounting professionals need to embrace innovation and develop digital literacy as Industry 4.0 and its subsequent evolution into Industry 5.0 take shape to be relevant in a business environment that is becoming more dynamic and interconnected by the day. In this era, accountants may create sustainable value creation and contribute to the long-term success of organizations by reinventing established accounting processes and utilizing the potential of emerging technology. Thus, it is crucial to understand the consequences of Industry 4.0 and 5.0 for a variety of fields, including education, as its influence grows and changes over time. This chapter examines the relationship between soft technology disruption and accounting education and looks ahead to Industry 5.0. Attempts have been made to analyze the awareness levels among students and professionals about this paradigm shift along with its impact on accounting education. Along with, the study measures the impact of soft technologies on organizational efficiency and students' readiness to adapt to these changes.

### 10.1.1 Problem Statement

The corporate landscape is changing due to Industry 4.0 and its progression towards Industry 5.0, leading to both unprecedented opportunities and difficulties for the accounting profession. To ensure relevance and effectiveness in the digital era, accounting education and practices need to be re-evaluated in light of the integration of modern technology, changing market dynamics, and increasing stakeholder expectations. For educators, practitioners, and policymakers alike, managing the complexities of soft technology disruption and making the shift to accounting education and practices aligned with Industry 4.0 and 5.0 is a challenge that needs to be met to maintain relevance in the digital era.

### 10.1.2 Research Questions

- What is the level of awareness about Industry 4.0 technologies, such as AI, big data analytics, blockchain, and automation, among Accounting Professionals and Students?
- How does soft technology disruption, driven by Industry 4.0 and evolving towards Industry 5.0, influence Accounting Education?
- How does the convergence of Industry 4.0 and 5.0 impact Organizational Performance?
- How does adopting Industry 4.0 and 5.0 Technologies in the Accounting Curriculum impact students' readiness?

## 10.2 LITERATURE REVIEW

### 10.2.1 Integration of Disruptive Technologies in Accounting Practices

Accounting procedures have undergone significant changes as a result of Industry 4.0, which is defined by the integration of cyber-physical systems, the IoT, blockchain, AI, and big data analytics [7]. Automation, real-time data processing, and advanced analytics are made possible by these technologies, which simplify the processes involved in financial reporting, auditing, and decision-making [8]. IoT makes it easier to gather and track financial data continuously, which improves the timeliness and accuracy of financial reporting [9]. AI algorithms enhance decision-making and risk management by executing intricate data analysis, identifying abnormalities, and forecasting financial trends [10]. Blockchain reduces fraud and enhances audit trails by ensuring security and transparency in financial transactions [11]. Big data analytics in accounting makes it possible to evaluate enormous volumes of financial data and derive insightful information that helps with performance assessment and strategy planning [12].

Technological advancement emphasizes human-centric technology to foster innovation and collaboration in accounting procedures, such as wearables and augmented reality (AR) [12]. The skill sets and capabilities of accounting professionals need to change to make this transformation. The integration of AR in accounting can enhance comprehension and decision-making by enabling more dynamic and immersive data visualization and analysis.

Digital industrial revolution demands expertise in digital literacy, data analysis, and technology-driven problem solving. Alongside technical competencies, soft skills like communication, creativity, and adaptability are becoming more and more important [13]. Accounting professionals may find the quick speed of technology change intimidating, as it necessitates constant learning and adjustment [14]. Curriculum needs to be updated, since developing technology and its applications may not be adequately covered in traditional accounting education [14]. There are plenty of chances to improve efficiency despite these obstacles. Professionals can concentrate on strategic operations by using automation and advanced analytics to reduce the time and effort needed for mundane accounting processes [15]. Better business outcomes are supported by real-time data processing and predictive analytics, which improve the caliber of financial decisions [16]. Blockchain technology increases confidence and compliance by ensuring safe and transparent financial transactions [11].

The convergence of big data analytics, blockchain, AI, and cyber-physical systems, or in other words paradigm of Industry 4.0, has significantly improved organizational performance. Real-time decision-making, enhanced data accuracy, and automation are all made possible by these technologies. Automation boosts productivity and operational efficiency by lowering human labor and mistake rates [17]. IoT ensures gathering and monitoring of real-time data, delivering timely insights that improve operational responsiveness and agility [18]. Deep insights into data patterns are provided by AI-driven analytics, enhancing competitive advantage and assisting in strategic decision-making [19]. Blockchain technology guarantees that transactions are safe and transparent, which lowers fraud and boosts confidence in financial procedures [20]. Strategic planning and performance evaluation are supported by big data analytics, which processes massive amounts of information to derive useful insights [21].

Automation and AI are incorporated into accounting programs to aid students in gaining vital technical skills. The significance of students comprehending and collaborating with AI systems that can learn, reason, and adapt [22]. Students need to be more equipped to handle automated procedures and data analytics, which are essential for today's accounting jobs as big data analytics augments students' capacity to construe extensive datasets, elevating their aptitude for decision-making [23]. Students improve their analytical skills via hands-on experience with big data tools and approaches. Students are encouraged to adopt a lifelong learning perspective by the requirement that the accounting curriculum be continuously modified to incorporate Industry 4.0 technologies.

According to Surianti [24], an adaptive curriculum makes it easier for pupils to keep up with technology developments, which improves their preparedness for obstacles they may face at the workplace. According to Supriadi [25], creating adaptable learning paths that include new technology guarantees that students may swiftly become accustomed to new resources and techniques, preserving their marketability.

## 10.3 CONCEPTUAL FRAMEWORK

### 10.3.1 Impact of Soft Technology Disruption on Accounting Education

Digitalization and automation, or 'soft technology disruption', radically alters educational paradigms and conventional accounting procedures. Disruptive developments like cyber-physical systems and human–machine collaboration are brought about by the digital revolution, and this calls for a thorough review of accounting education frameworks to make sure graduates have the necessary skills for the digital age, accounting programs must incorporate cutting-edge technology like blockchain and AI. To meet industry demands and promote technological fluency among accounting professionals, curriculum modifications should place a higher priority on practical training in data analytics, software applications, and cybersecurity measures [15, 26].

Industry 4.0 and 5.0 emphasize how crucial it is for accounting professionals to be flexible and always learning new things. Critical thinking, problem-solving, and communication skills must be given top priority in accounting education so that graduates can successfully navigate changing regulatory frameworks and dynamic technological environments. Furthermore, it is imperative to prioritize multidisciplinary cooperation and moral decision-making to develop well-rounded accounting professionals who can handle challenging situations in the digital era [27, 28].

H1*: Accounting professionals and trainee students are significantly aware of Industry 4.0 technologies.*

H2: *Soft technology disruption, particularly within the context of Industry 4.0 and 5.0, significantly impacts accounting education.*

### 10.3.2 Convergence of Industry 4.0 and Organizational Performance

Organizational agility, efficiency, and risk management capacities are improved when Industry 4.0 technologies and accounting processes come together [29]. Organizations can attain operational excellence and continuous improvement in accounting processes by integrating Industry 4.0 and Lean Six Sigma approaches [30]. Supply chain risk analytics are facilitated by digital technology and Industry 4.0, which also enhance supply chain resilience and allow for proactive risk mitigation techniques [31]. Traditional accounting procedures are being transformed by the use of new-age technologies [32].

Accounting professionals can streamline financial reporting, auditing, and decision-making processes with the use of these technologies, which enable automation, real-time data processing, and advanced analytics [8]. To improve cooperation and creativity in accounting procedures, Industry 5.0 places even more emphasis on integrating human-centric technologies, such as AR. The adoption of Industry 4.0 has resulted in a need for accounting professionals to reassess their skill sets, with a focus on digital literacy, data analysis, and technology-driven problem-solving [33]. To successfully negotiate the intricacies of Industry 4.0 and 5.0, soft skills like communication, creativity, and adaptability are becoming more and more important in addition to technical competencies [13]

H3: *The integration of disruptive technologies in accounting practices, aligned with Industry 4.0 and 5.0 paradigms, enhances organizational performance.*

### 10.3.3 Industry 4.0 Technologies in Accounting Curriculum and Students' Readiness

Accounting curricula that integrate Industry 4.0 technologies—such as blockchain, AI, big data analytics, and robotic process automation—improve students' technical proficiency and get them ready for the profession's impending digital transformation [34]. Students can obtain practical experience with technology routinely used in modern accounting practices by focusing on the usage of advanced software and tools in accounting education. Students are better prepared to handle the complexity of the Industry 4.0 and 5.0 landscape by receiving training in cutting-edge technology unique to the accounting and auditing industry, such as FinTech apps and AI-based auditing tools [34].

It is ensured that accounting curricula are relevant and aligned with the changing needs of accounting professionals by routinely revising them to reflect industry developments and regulatory changes [35].

*H4: Adapting digital technologies in accounting curriculum significantly enhances students' readiness for the Industry 4.0 and 5.0 era.*

## 10.4 RESEARCH METHODOLOGY

### 10.4.1 Research Design

To investigate the effects of incorporating Industry 4.0 technologies into the accounting curriculum on students' preparedness for professional positions, this study uses a quantitative research approach. Information has been gathered from accounting professionals and trainee students using a semi-structured survey instrument developed upon a 5-point Likert scale.

### 10.4.2 Sampling Method

Participants who were either accounting professionals or student trainees working under professionals and pursuing accounting qualifications were chosen using purposeful sampling. The aim was to make sure that the sample included people who had relevant experience and knowledge of the topic.

### 10.4.3 Survey Instrument

The semi-structured survey instrument has been applied. The reliability of the instrument has been well established using alpha values, which for every statement have been greater than 0.80. Validity issues were addressed with the help of senior researchers in the field.

### 10.4.4 Demographic Information

Basic details about the respondents were gathered in this part, such as gender, age, and their current title.

### 10.4.5 Technical Awareness

This section evaluated respondents' knowledge and comfort levels with a range of Industry 4.0 technologies, including automation, big data analytics, blockchain, and AI.

### 10.4.6 Impact of Technology Disruption on Accounting Education

This section focused on understanding the perception of respondents about the impact of soft technologies on 'Accounting Education'.

### 10.4.7 Curriculum Adaptation

This section focused on respondents' perceptions of how well the present accounting curriculum integrates these technologies.

### 10.4.8 Professional Readiness

Statements included in this part evaluated the perceived readiness of students to enter the professional accounting field after being exposed to Industry 4.0 and 5.0 technologies.

**TABLE 10.1**
**Demographic profile**

| Profile | Attribute | Number of respondents | Percentage |
|---|---|---|---|
| Gender | Male | 175 | 70 |
| | Female | 75 | 30 |
| Age | Below 30 years | 110 | 44 |
| | 30–45 years | 86 | 34.4 |
| | 45–60 years | 34 | 13.6 |
| | 60 Years and above | 20 | 8 |
| Title | Students | 97 | 38.8 |
| | CA | 73 | 29.2 |
| | CMA | 42 | 16.8 |
| | CS | 11 | 4.4 |
| | ICAI | 27 | 10.8 |

*Source:* Researcher's Compilation
*Abbreviations:* CA, chartered accountant; CMA, cost and management accountant; CS, company secretaries; ICAI, Institute of Chartered Accountants of India.

## 10.5 DATA ANALYSIS

The collected data has been analyzed using descriptive statistics and values of $t$ statistic and hypotheses have been tested based upon $p$-value derived (Table 10.1).

Male respondents made up 70% of the total participants in the survey sample, while female respondents made up 30% (Table 10.1). The gender group's breakdown emphasizes the predominant role of males in the accounting field. Based on the age distribution, it can be observed that 44% of the sample's respondents were under 30 years old. The age group of 30–45 years old, which comprised 34.4% of the respondents, comes next. Respondents 60 years of age and older made up 8% of the sample, while those between 45 and 60 years old made up 13.6%. This suggests a young demographic, with a sizable percentage of participants being in their early professional years. The respondents' professional titles show a wide variety of accounting-related specializations. Students pursuing accounting education make up 38.8% of the sample, indicating a substantial representation from academia. The proportion of respondents who were chartered accountants, at 29.2%, show their significance in the professional accounting industry. Company secretaries made up 4.4% and cost and management accountants 16.8%. Respondents of 10.8% were members of the Institute of Chartered Accountants of India.

### 10.5.1 Descriptive Analysis and Hypothesis Testing

There are notable disparities between accounting professionals and trainee students in their awareness of Industry 4.0 and prospective Industry 5.0 technologies, according to the investigation done on the subject (Table 10.2). Accounting professionals show

**TABLE 10.2**
**Awareness levels among accounting professionals and students**

| Factors | *N* | df | Mean | SD | *t*-Value | *p*-Value |
|---|---|---|---|---|---|---|
| AL1 | 250 | 249 | 2.84 | 0.55 | 16.09 | 0.001 |
| AL2 | 250 | 249 | 2.93 | 0.50 | 22.29 | 0.001 |
| AL3 | 250 | 249 | 3.01 | 0.72 | 17.75 | 0.001 |
| Overall | 250 | 249 | 2.96 | 0.50 | 21.26 | 0.001 |

*Source:* Researcher's CompilationTop of Form

consistently greater levels of awareness than training students across three different parameters assessing awareness (AL1, AL2, and AL3). In particular, trainee students had slightly lower mean scores across all variables, whereas accounting professionals have mean awareness scores of 2.84, 2.93, and 3.01 for factors AL1, AL2, and AL3, respectively. As *t*-values range between 16.09 and 22.29, with considerably low *p*-values of 0.001, the directional hypothesis stands accepted.

H1*: Accounting professionals and trainee students are significantly aware of Industry 4.0 technologies*

Though the directional hypothesis has been accepted, compared to trainee students, accounting professionals are reflected to have a more thorough understanding of Industry 4.0 technology. A moderate level of awareness is highlighted by both groups. Combined overall mean awareness score is 2.96, which results from combining the three variables. The differences show that focused educational programs are necessary to close the knowledge gap between accounting experts and trainee students and ensure that the profession has a more consistent understanding of Industry 4.0 technology. The *p*-value is less than 0.05; thus, our first directional hypothesis has been accepted. But there is a sign that the awareness level among students of soft technologies being used in accounting education is lower than that of professionals.

Strong findings emerge from the examination of comments assessing how soft technology disruption affects accounting education, especially in Industry 4.0 and 5.0. Soft technological disruption appears to have a consistent and statistically significant impact on accounting education across all examined aspects, including statements STAE1, STAE2, and STAE3 (Table 10.3). For statement STAE1, which evaluates a cognitive aspect of soft technology disruption, the average score is 3.60 (SD = 0.650), showing a significant impact. Similarly, statement STAE3 shows a mean score of 4.00 (SD = 0.62), showing that soft technology disruption has a significant impact on accounting education. *p*-Value being less than 0.05, the second directional hypothesis has been accepted.

H2: *Soft technology disruption, particularly within the context of Industry 4.0 and 5.0, significantly affects accounting education.*

This impact is widespread, as evidenced by the whole analysis, which has an aggregate mean score of 3.81 (SD = 0.60). This shows a steady trend in several areas, emphasizing the significant effects of soft technology disruption on the development of accounting education. As a result, our results offer empirical backing for

**TABLE 10.3**
**Soft technology disruption and accounting education**

| Statements | *N* | df | Mean | SD | *t*-Value | *p*-Value |
|---|---|---|---|---|---|---|
| STAE1 | 250 | 249 | 3.60 | 0.65 | 18.500 | 0.001 |
| STAE2 | 250 | 249 | 3.82 | 0.80 | 16.200 | 0.001 |
| STAE3 | 250 | 249 | 4.00 | 0.62 | 22.700 | 0.001 |
| Overall | 250 | 249 | 3.81 | 0.60 | 20.000 | 0.001 |

*Source:* Researcher's Compilation

**TABLE 10.4**
**Disruptive technologies and organizational performance**

| Statements | *N* | df | Mean | SD | *t*-Value | *p*-Value |
|---|---|---|---|---|---|---|
| OP1 | 250 | 249 | 4.10 | 0.85 | 21.50 | 0.001 |
| OP2 | 250 | 249 | 3.90 | 0.75 | 18.20 | 0.001 |
| OP3 | 250 | 249 | 3.80 | 0.80 | 20.10 | 0.001 |
| Overall | 250 | 249 | 3.93 | 0.78 | 19.80 | 0.001 |

*Source:* Researcher's Compilation

the directional hypothesis, showing that, in the context of Industry 4.0 and 5.0, soft technology disruption has a notable impact on accounting education.

Statements evaluating the adoption of technologies in accounting processes exhibit consistent and statistically significant results. Results point to a robust and positive relationship between technologies and organizational performance across all measured dimensions, which include statements OP1, OP2, and OP3 (Table 10.4). Statement OP1 has a mean score of 4.10 (SD = 0.85) and focuses on the impact of technology integration on an organization's productive capacity. Similarly, mean scores of 3.90 (SD = 0.750) and 3.80 (SD = 0.800) for statements OP2 and OP3, respectively, support the idea that technologies play a major role in the success of organizations. An aggregated mean score of 3.93 (SD = 0.780) suggests a steady trend across various dimensions. The overall impact further supports these findings. Thus, our findings offer strong empirical support for the hypothesis that technology integration improves organizational performance.

After evaluation of each statement (SR1, SR2, SR3, and SR4), the mean scores show firm agreement with the beneficial effects of these curricular modifications (Table 10.5). The results indicate that respondents believe there has been a significant positive impact of including digital technology in the curriculum. The overall mean score of 3.90 (SD = 0.50) confirms a favorable influence of digital technologies on students' preparation for the Industry 4.0 and 5.0 era. All four statements exhibit high *t*-values and extremely low *p*-values (0.001) supporting the statistical significance of these results, which have a very low possibility of occurring by accident. The *p*-value

**TABLE 10.5**
**Digital technologies and students' readiness for the Industry 4.0 and 5.0**

| Statements | N | df | Mean | SD | t-Value | p-Value |
|---|---|---|---|---|---|---|
| SR1 | 250 | 249 | 3.70 | 0.70 | 19.28 | 0.001 |
| SR2 | 250 | 249 | 4.05 | 0.550 | 34.09 | 0.001 |
| SR3 | 250 | 249 | 3.85 | 0.75 | 21.33 | 0.001 |
| SR4 | 250 | 249 | 4.00 | 0.60 | 31.25 | 0.001 |
| Overall | 250 | 249 | 3.90 | 0.50 | 35.00 | 0.001 |

*Source:* Researcher's Compilation

is less than 0.05 for all the statements of the fourth directional hypothesis, and thus, it is accepted.

*H4: Adapting digital technologies in the accounting curriculum significantly enhances students' readiness for the Industry 4.0 and 5.0 era.*

## 10.6 DISCUSSION

The study of four hypotheses about different facets of technology awareness and its effects on accounting education and organizational performance is summarized in the Discussion section, along with its consequences. This chapter discusses the literature's theoretical contributions and the practical ramifications of accounting stakeholders.

### 10.6.1 Theoretical Implications

#### 10.6.1.1.1 Awareness of Industry 4.0 Technologies

The analysis emphasizes the importance of awareness about the advancement of technologies in the field of 'Accounting'. The study shows that accounting professionals and trainee students have quite different levels of awareness when it comes to Industry 4.0 technologies [25, 36]. This study adds to the body of literature by emphasizing how organizational culture, experience, and education shape accounting professionals' understanding of technology [23].

#### 10.6.1.1.2 Impact of Soft Technology Disruption on Accounting Education

The idea that soft technology disruption has a substantial impact on accounting education and requires curriculum adaptation to include issues related to Industry 4.0 and 5.0 paradigms is supported by empirical evidence [24, 36]. To investigate ways to improve accounting education and better equip students for the difficulties presented by technologies, the study adds value to the theoretical models of educational innovation and change management [37].

#### 10.6.1.1.3 Integration of Technologies and Organizational Performance

According to the investigation, there is a direct correlation between organizational success in the accounting context and incorporating technologies [18, 21]. This result emphasizes the potential for technology adoption to increase productivity, efficiency, and competitiveness, which is consistent with theories of technological innovation and organizational transformation [17, 38].

#### 10.6.1.1.4 Digital Technologies in Accounting Curriculum and Students' Readiness

This study enriches educational innovation theories by illustrating that incorporating technologies such as data analytics, AI, blockchain, the IoT, and AR yields tangible benefits (Surianti, 2020; Mian et al., 2020). These findings support the need for continuous curriculum adaptation to keep pace with technological advancements [25, 37]. The study advocates competency-based education frameworks by highlighting the necessity for educational programs to equip students with practical skills and theoretical knowledge critical for Industry 4.0 and 5.0, ensuring that graduates are prepared to meet the growing demands of the industry [23, 39].

## 10.7 PRACTICAL IMPLICATIONS

### 10.7.1 Professional Development and Training

The significance of continuous professional development and training programs for students in the 'Accounting' domain is highlighted by differences in technology awareness [25, 37]. Partnerships between academic institutions and accounting businesses can support customized training initiatives designed to improve students' and professionals' technological literacy [37].

### 10.7.2 Curricular Reform in Accounting Education

This study highlights the necessity of curriculum revision in accounting education to meet the demands of the digital age [24, 39]. Incorporating modules focused on cybersecurity, data analytics, and emerging technologies can guarantee that graduates possess the abilities required for the technology-driven accounting industry [36].

### 10.7.3 Strategic Technology Adoption in Accounting Firms

The study's conclusions can help accounting companies make strategic decisions about implementing technology, which will improve client service and operational effectiveness [18, 38].

### 10.7.4 Policy and Regulation

Regulatory agencies and policymakers can use the results of studies to encourage technology innovation and adoption in the accounting industry [40, 41]. While reducing risks and obstacles, supportive policies for cybersecurity, technological infrastructure, and research and development can promote digital transformation [42].

## 10.8 FUTURE RESEARCH

Future research is suggested in the form of longitudinal studies on technology adoption in the field of accounting education; the impact of specific technologies needs to be explored too. A detailed skill gap analysis is also required to understand and integrate technological advancements in accounting.

## 10.9 CONCLUSION

Technology's advancement, especially the Industry 4.0 and 5.0 paradigms, has significantly reshaped the accounting industry. The current investigation has examined the effects of technology progress on several aspects of the accounting domain, providing theoretical understandings and pragmatic recommendations for interested parties. The study's conclusions highlight how important it is for accounting professionals and prospective students to be aware of technology. Variations in the degree of awareness were noted, highlighting the necessity of focused educational programs to close the knowledge gap and guarantee a more consistent comprehension of Industry 4.0 technologies throughout the industry. The study also provides empirical evidence to substantiate the claim that soft technology disruption has a big influence on accounting education. Accounting firms can achieve success in the field of organizational performance by integrating disruptive technology. Last but not least, the study highlights that if the students of accounting are well-oriented in digital technologies, then their adaptability to Industry 4.0 and 5.0 paradigms is enhanced.

## REFERENCES

1. Bueno, Adauto, Moacir Godinho Filho, and Alejandro G. Frank. "Smart production planning and control in the Industry 4.0 context: A systematic literature review." *Computers & Industrial Engineering* 149 (2020): 106774. https://doi.org/10.1016/j.cie.2020.106774
2. Lu, Yang. "Industry 4.0: A survey on technologies, applications and open research issues." *Journal of Industrial Information Integration* 6 (2017): 1–10. https://doi.org/10.1016/j.jii.2017.04.005
3. Ghobakhloo, M. "Industry 4.0, digitization, and opportunities for sustainability." *Journal of Cleaner Production* 252 (2020): 119869. https://doi.org/10.1016/j.jclepro.2019.119869.
4. Jamwal, Anurag, R. Agrawal, M. Sharma, and A. Giallanza. "Industry 4.0 technologies for manufacturing sustainability: A systematic review and future research directions." *Applied Sciences* 11, no. 12 (2021): 5725. https://doi.org/10.3390/app11125725.
5. Ghani, E. K., A. Ilias, K. Muhammad, N. A. Rosley, M. M. Ali, and C. Sukmadilaga. "Expectation gap on accounting graduates' Skill attributes between accounting educators and employers in Industry 4.0 environment: A Malaysian evidence." *Educational Administration: Theory and Practice* 30, no. 3 (2024): 1–16. https://doi.org/10.52152/kuey.v30i3.1059.
6. Javaid, Muhammad, A. Haleem, R. P. Singh, R. Suman, and E. S. Gonzalez. "Understanding the adoption of Industry 4.0 technologies in improving environmental sustainability." *Sustainable Operations and Computers* 3 (2022): 203–217. https://doi.org/10.1016/j.susoc.2022.01.008.

7. Sharma, Anil K., R. Bhandari, C. Pinca-Bretotean, C. Sharma, S. K. Dhakad, and A. Mathur. "A study of trends and industrial prospects of Industry 4.0." *Materials Today: Proceedings* 47 (2021): 2364–2369. https://doi.org/10.1016/j.matpr.2021.04.321.
8. Margherita, Antonio, and Andrea M. Braccini. "The role of advanced analytics in improving accounting efficiency." *Journal of Business Research* 130 (2024): 413–426.
9. Rabbani, Muhammad R. "Impact of digital advancements on accounting, auditing and reporting literature: Insights, practice implications and future research directions." *Journal of Accounting & Organizational Change* (2024). https://doi.org/10.1108/JAOC-01-2024-0028.
10. Sharma, Anil, and Dinesh K. Jain. "Development of Industry 4.0." In A. Nayyar & A. Kumar (Eds.), *A Roadmap to Industry 4.0: Smart Production, Sharp Business and Sustainable Development* (pp. 23–38). Cham: Springer, 2020. https://doi.org/10.1007/978-3-030-14544-6_2.
11. Olsen, Christian. "Toward a digital sustainability reporting framework in organizations in the Industry 5.0 Era: An accounting perspective." In *European, Asian, Middle Eastern, North African Conference on Management & Information Systems* (pp. 463–473). Cham: Springer International Publishing, 2022. https://doi.org/10.1007/978-3-031-17746-0_37.
12. Maddikunta, P. K. R., Quoc V. Pham, B. Prabadevi, N. Deepa, K. Dev, T. R. Gadekallu, and M. Liyanage. "Industry 5.0: A survey on enabling technologies and potential applications." *Journal of Industrial Information Integration* 26 (2022): 100257. https://doi.org/10.1016/j.jii.2021.100257.
13. Varma, Ashish, and Anuj Sharma. "Assessing the impact of accounting analytic capabilities on firm performance." *Journal of Computer Information Systems* (2024): 1–15. https://doi.org/10.1080/08874417.2024.2329137
14. Tavares, M. C., G. Azevedo, R. P. Marques, and M. A. Bastos. "Challenges of education in the accounting profession in the era 5.0: A systematic review." *Cogent Business & Management* 10, no. 2 (2023): 2220198. https://doi.org/10.1080/23311975.2023.2220198
15. Mariani, Marco, and Daniele Borghi. "Educational paradigms and Industry 4.0." *Education and Information Technologies* 24, no. 4 (2019): 2341–2358. https://doi.org/10.1016/j.techfore.2019.119752.
16. Sharma, Anil K., R. Bhandari, C. Pinca-Bretotean, C. Sharma, S. K. Dhakad, and A. Mathur. "A study of trends and industrial prospects of Industry 4.0." *Materials Today: Proceedings* 47 (2021): 2364–2369. https://doi.org/10.1016/j.matpr.2021.04.321.
17. Tortorella, Glaucio L., Tiago A. Saurin, Flavio S. Fogliatto, Diego Tlapa Mendoza, Javier Moyano-Fuentes, Paolo Gaiardelli, and Enrique Macias de Anda. "Digitalization of maintenance: Exploratory study on the adoption of Industry 4.0 technologies and total productive maintenance practices." *Production Planning & Contro*l 35, no. 4 (2024): 352–372. https://doi.org/10.1080/09537287.2022.2083996
18. Marcucci, Giulio, Sara Antomarioni, Filippo Emanuele Ciarapica, and Maurizio Bevilacqua. "The impact of operations and IT-related Industry 4.0 key technologies on organizational resilience." *Production Planning & Control* 33, no. 15 (2022): 1417–1431. https://doi.org/10.1080/09537287.2021.1874702
19. Elgendy, Nabil, Ahmed Elragal, and Tero Päivärinta. "DECAS: A modern data-driven decision theory for big data and analytics." *Journal of Decision Systems* 31, no. 4 (2022): 337–373. https://doi.org/10.1080/12460125.2021.1894674.
20. Pimentel, Elio, and Emmanuel Boulianne. "Blockchain in accounting research and practice: Current trends and future opportunities." *Accounting Perspectives* 19, no. 4 (2020): 325–361. https://doi.org/10.1111/1911-3838.12239.
21. Frederico, Gabriel F. "From Supply Chain 4.0 to Supply Chain 5.0: Findings from a systematic literature review and research directions." *Logistics* 5, no. 3 (2021): 49. https://doi.org/10.3390/logistics5030049.

22. Damerji, Hamza, and Ahmed Salimi. "Mediating effect of use perceptions on technology readiness and adoption of artificial intelligence in accounting." *Accounting Education* 30, no. 2 (2021): 107–130. https://doi.org/10.1080/09639284.2021.1872035.
23. Prakash, C., Yadav, R., Dangi, A., & Singh, A. (2023). Factor analysis approach to study mobile applications' characteristics and consumers' attitudes. In P. K. Gupta, R. Singhal, P. Jain, D. Kumar, & P. K. Shukla (Eds.), *Computational Intelligence for Engineering and Management Applications: Select Proceedings of CIEMA 2022* (pp. 783–798). Singapore: Springer Nature Singapore. https://dokumen.pub/computational-intelligence-select-proceedings-of-incite-2022-9811973458-9789811973451.html
24. Surianti, Meily. "Development of accounting curriculum model based on industrial revolution approach." *Development* 11, no. 2 (2020). https://core.ac.uk/download/pdf/287193003.pdf
25. Supriadi, I., K. A. Rahardjo, and M. D. Suprihandari. "Expectations and standards for accounting professions and implications in the learning pattern in Industrial 4.0." *Ilomata International Journal of Tax and Accounting* 1, no. 2 (2020): 66–73. https://doi.org/10.52728/ijtc.v1i2.58.
26. Erro-Garcés, A., and Miguel E. Aramendia-Muneta. "The role of human resource management practices on the results of digitalisation. From Industry 4.0 to Industry 5.0." *Journal of Organizational Change Management* 36, no. 4 (2023): 585–602. https://doi.org/10.1108/JOCM-11-2021-0354.
27. Arkhipova, D., Maria Montemari, Carla Mio, and Salvatore Marasca. "Digital technologies and the evolution of the management accounting profession: A grounded theory literature review." *Meditari Accountancy Research* 32, no. 7 (2024): 56–85. https://doi.org/10.1108/MEDAR-07-2023-2097.
28. Belinski, Rafael, Alexandre M. Peixe, Gabriel F. Frederico, and Julio A. Garza-Reyes. "Organizational learning and Industry 4.0: Findings from a systematic literature review and research agenda." *Benchmarking: An International Journal* 27, no. 8 (2020): 2435–2457. https://doi.org/10.1108/BIJ-04-2020-0158.
29. Saini, C. P., Nair, M. K., & Shankar, K. T. (2020). Recycling and reuse of fashion: A way to sustainable environment. In S. S. Muthu & M. A. Gardetti (Eds.), *Circular Economy and Re-Commerce in the Fashion Industry* (pp. 47–53). Hershey, PA: IGI Global. https://doi.org/10.4018/978-1-7998-5116.
30. Skalli, Djamel, A. Cherrafi, A. Charkaoui, A. Chiarini, A. Shokri, J. Antony, and M. Foster. "Integrating lean six sigma and Industry 4.0: Developing a design science research-based LSS4.0 framework for operational excellence." *Production Planning & Control* (2024): 1–27. https://doi.org/10.1080/09537287.2024.2341698.
31. Ivanov, Dmitry, Chris S. Tang, Alain Dolgui, Daniele Battini, and Arij Das. "Researchers' perspectives on Industry 4.0: Multi-disciplinary analysis and opportunities for operations management." *International Journal of Production Research* 59, no. 7 (2021): 2055–2078. https://doi.org/10.1080/00207543.2020.1798035.
32. Sharma, A., & Jain, D. K. (2020). Development of industry 4.0. In A. Yadav, M. K. Tiwari, & S. K. Bhatia (Eds.), *A Roadmap to Industry 4.0: Smart Production, Sharp Business and Sustainable Development* (pp. 23–38). Cham: Springer. https://doi.org/10.1007/978-3-030-14544-6_2.
33. Bongomin, Ocident, Gilbert Gilibrays Ocen, Eric Oyondi Nganyi, Alex Musinguzi, and Timothy Omara. "Exponential disruptive technologies and the required skills of industry 4.0." *Journal of Engineering* 2020, no. 1 (2020): 4280156. https://doi.org/10.1155/2020/4280156
34. Thottoli, M. M. "Leveraging Information Communication Technology (ICT) and Artificial Intelligence (AI) to enhance auditing practices." *Accounting Research Journal* 37, no. 2 (2024): 134–150. https://doi.org/10.1108/ARJ-09-2023-0269.

35. Chandler, Nick. "Shifting work competencies in an emerging economy: A comparison of accounting students' perceptions of demand and supply." *Higher Education, Skills and Work-Based Learning* (2024). https://doi.org/10.1108/HESWBL-08-2023-0213.
36. Mian, Shahid H., Muhammad A. Gulzar, and Muhammad Anwar. "Awareness of Industry 4.0 technologies among accounting professionals and students." *Journal of Accounting and Finance* 60, no. 3 (2020): 113–126. https://doi.org/10.3390/su12156100.
37. Tsiligiris, Vasilis, and David Bowyer. "Exploring the impact of 4IR on skills and personal qualities for future accountants: A proposed conceptual framework for university accounting education." *Accounting Education* 30, no. 6 (2021): 621–649. https://doi.org/10.1080/09639284.2021.1938616
38. Pimentel, Elio, and Emmanuel Boulianne. "Blockchain in accounting research and practice: Current trends and future opportunities." *Accounting Perspectives* 19, no. 4 (2020): 325–361. https://doi.org/10.1111/1911-3838.12239.
39. do Rosário Cabrita, Maria, Hossein Safari, and M. D. P. M. Dueñas. "Preparing for education 4.0: Skills facing economic, social and environmental challenge." *International Journal of Innovation, Management and Technology* 11, no. 1 (2020): 33–37.
40. Mubarak, Muhammad F., S. Tiwari, Margarita Petraite, Muhammad Mubarik, and Raja Mohd Rasi. "How Industry 4.0 technologies and open innovation can improve green innovation performance?" *Management of Environmental Quality: An International Journal* 32, no. 5 (2021): 1007–1022. https://doi.org/10.1108/MEQ-11-2020-0266.
41. Prakash, C., R. Yadav, A. Singh, and K. Yadav. "Modeling the entrepreneurial intention among business students." *Pacific Business Review International* 14, no. 3 (2021): 77–90.
42. Popkova, Elena G., and Bruno S. Sergi. "Human capital and AI in Industry 4.0. convergence and divergence in social entrepreneurship in Russia." *Journal of Intellectual Capital* 21, no. 4 (2020): 565–581. https://doi.org/10.1108/JIC-09-2019-0224.

# Appendix 'A'

## SECTION A

**1. Survey Instrument**

1. **NAME:** ____________________
2. **GENDER**:

MALE ☐ FEMALE ☐

**2. Age**

- Below 30 years
- 30–45 years
- 45–60 years
- 60 years and above

**3. Professional Title**

- Student
- Chartered accountant
- Cost and management accountant
- Company secretary
- Member of the Institute of Chartered Accountants of India (ICAI)

## SECTION-B

Rate the following statements on a scale of 1–5 (where Strongly Disagree = 1, Disagree = 2, Neutral = 3, Agree = 4, Strongly Agree = 5)

| Factor | Statement | 1 | 2 | 3 | 4 | 5 |
|---|---|---|---|---|---|---|
| (Technology Awareness) AL1 | I am aware of the concept of Industry 4.0. | | | | | |
| AL2 | I am knowledgeable about how Industry 4.0 technologies can be applied in accounting. | | | | | |
| AL3 | I am aware of how Industry 4.0 technologies impact the accounting profession | | | | | |

*(Continued)*

| Factor | Statement | 1 | 2 | 3 | 4 | 5 |
|---|---|---|---|---|---|---|
| (Technology Disruption) STAE1 | Learning through digital platforms has improved my cognitive skills in accounting | | | | | |
| STAE2 | Using soft technologies has prepared me better for real-world accounting tasks | | | | | |
| STAE3 | The integration of soft technologies has made learning accounting more effective and engaging | | | | | |
| (Organization Productivity) OP1 | The integration of disruptive technologies has significantly increased our organization's productive capacity | | | | | |
| OP2 | Advanced technologies have streamlined our accounting processes, leading to better efficiency | | | | | |
| OP3 | Our organization has achieved higher levels of performance and success due to the integration of disruptive technologies | | | | | |
| (Student Readiness) SR1 | The curriculum has effectively equipped me with the necessary technical skills for Industry 4.0 and 5.0. | | | | | |
| SR2 | I have a solid grasp of the theoretical concepts behind the digital tools used in the accounting profession | | | | | |
| SR3 | I believe I am ready to meet the expectations of future employers in the digital era | | | | | |
| SR4 | I am confident in using digital technologies to solve real-world accounting problems | | | | | |

# 11 The Trend and Type Analysis of Various Disruptive Technologies in Healthcare

*Muddapu Raja, T. N. V. R. L. Swamy, and Chand Prakash*

## 11.1 INTRODUCTION

Disruptive technology is the use of technology that influences the functions of an industry [4]. It replaces traditional ways of doing things and provides users with an alternative or new way to obtain goods and services efficiently [39]. These technologies include 3D printing, artificial intelligence, augmented reality, cloud computing, big data, blockchain technology, and more. Technology disruption has emerged in several industries, such as education, agriculture, healthcare, energy, and logistics [5].

The innovation of technology in healthcare results in a new concept called Healthcare 4.0 that enables real-time customization of healthcare for caretakers, professionals, and patients. It shifts organizations from hospital to patient-centric and provides the best patient healthcare results [41]. In Healthcare 4.0, some of the technologies used in the medical field are big data, the internet of things (IoT), cloud computing [13], electronic health records [37], artificial intelligence, clinical decision support system (CDSS) [4], and blockchain [11].

Global healthcare big data analytics is a vast market divided into clinical, financial, and operational. In 2016, the market size of clinical was 1.65, financial was 2.38, and operational was 0.65 billion USD. It forecasted that by the end of 2025, the market would grow to 40% with clinical 11.35, financial 13.14, and operational 4.45 billion USD [40]. The IoT is also essential in medicine. A research report states that IoT will achieve a compound annual growth rate (CAGR) of 29.9% from 2019 and reach 322.2 billion USD by 2025 [26]. On 14th October 2022, Coherent Market Analysis released a press report, forecasting the current market share of global cloud computing in healthcare of 54.38 billion USD will increase to 15.6% of CAGR and reach 173.89 CAGR by 2030 [7]. Blockchain technology also has a considerable market share in healthcare. Precedence Research forecasts that by the end of 2032, this market will gain 34.02% of CAGR and reach 14.25 billion USD [29]. Simultaneously, this organization also forecasts that the 3.12 billion USD


DOI: 10.1201/9781032700953-13

market share of the Global CDSS market share will reach 6.9 billion USD by 2032 with a CAGR of 8.3% [30].

Based on the above observation of reports by various research organizations, the current chapter aimed to understand these five technologies, such as big data, IoT, cloud computing, blockchain, and CDSS, in healthcare and analyzed the current research trends of each technology. Additionally, this research also intended to perform categories of document analysis. The selection procedure of literature is common for the five technologies to overcome biasness. Table 11.1 shows the selection process of literature for annual production and document analysis.

## 11.2 DISRUPTIVE TECHNOLOGIES IN HEALTHCARE

### 11.2.1 Big Data

Big data starts with the requirement to analyze vast amounts of information for large companies like Google, Facebook, and Yahoo [45]. It is a continuous extraction of heterogeneous data in structured or unstructured form [22]. Simultaneously, it collects and analyzes previous work activities to provide insights for planning, designing, monitoring, and maintaining future works. It also helps reduce costs, improve efficiency, and increase work-related outcomes [24].

At first, big data represents the 3Vs: volume, velocity, and variety. Later, IBM and Microsoft added a "fourth V" Veracity [45]. Now it consists of 6Vs [18]: (1) Volume speaks about the amount of available data; (2) Variety represents the variation in the information and source, which is in the form of structured, unstructured, and semi-structured [35]; (3) Velocity deals with the movement, speed, and frequency of the data; (4) Veracity refers to the data reliability, quality, and relevance; (5) Variability deals with data consistency over period; and (6) Value deals with the worthiness of collected information [33].

When organizations effectively use big data, it helps them gain new insights, enhance value, and provide a competitive advantage [14]. Additionally, this data helps recommend goods and services to target audiences by analyzing consumer behavior [31]. In healthcare, the big data journey was started in 2013 with the report of McKinsey & Company. At present, it is serving in various purposes of healthcare, such as detection of fraud, prediction of epidemic spreading, medical devices' design, clinical, omics, patient care, pharmaceutical development, and more [34]. In addition to multiple benefits in healthcare, requirement of huge initial investment, lack of training and peoples' resistance to change, low quality of information [23], and privacy & security breaches make its implementation challenging [28].

#### 11.2.1.1 Annual Scientific Production of Big Data

The annual scientific production of documents shows the publication trend over a period that helps researchers understand the area's growth or decline. Using this information, a researcher can decide whether to conduct or change the research topic.

Figure 11.1 shows the annual scientific production of big data. It is showing literature from 2012 to 2024. However, as 2024 is the current year and only seven months have passed, comparing whole years of previous publications is

**TABLE 11.1**
**Filtration process of the literature**

| Technologies | Big data | | Blockchain | | Cloud computing | | Internet of things | | Clinical decision support system | |
|---|---|---|---|---|---|---|---|---|---|---|
| **Selection criteria** | **Include** | **Exclude** | **Include** | **exclude** | **Include** | **exclude** | **Include** | **exclude** | **Include** | **exclude** |
| **Database:** Scopus | 7,114 | – | 5,576 | – | 4,876 | – | 13,936 | – | 2,322 | – |
| **Search term:** “Name of technology” and “Healthcare” | | | | | | | | | | |
| **Document type:** Articles, review, books, book chapters, conference reviews & papers | 6,898 | 216 | 5,496 | 80 | 4,823 | 53 | 13,785 | 151 | 2,263 | 59 |
| **Language:** English | 6,723 | 175 | 5,469 | 27 | 4,799 | 24 | 13,715 | 70 | 2,220 | 43 |
| **Source type:** Journal, conference proceedings, book, book series | 6,702 | 21 | 5,452 | 17 | 4786 | 13 | 13,681 | 34 | 22,20 | 0 |

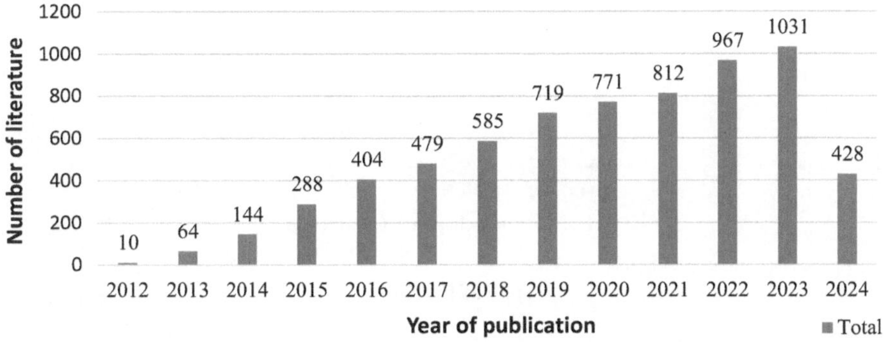


**FIGURE 11.1** Annual scientific production of big data

inappropriate. However, this projection of the current year's literature will help gain ideas for future publications. Nearly after the completion of 60% of the current year, only 428 publications are available, which is 41.5% of 2023. With this finding, the authors predict that there might be a slight chance of a reduction in the overall literature of 2024.

When observing the remaining years of literature, the bar chart shows that the publications in big data have been continuously booming since 2012, when only 10 publications were published. After 11 years, the publications reached 1,031, which is 100 times more than in 2012. Thus, after these observations, the authors suggested that future researchers can conduct research on big data and explore more to enhance current knowledge or provide solutions for potential issues.

#### 11.2.1.2 Big Data Document Category Analysis

The categories of documents are an analysis that differentiates types of published or ready-to-publish documents. This analysis will give researchers an idea about what type of research they can perform. There are several types of documents in the Scopus databases, such as articles, reviews, conference papers, notes, editorials, and more.

In this chapter, the authors considered six categories of documents: articles, books, book chapters, conference papers, conference reviews, and reviews.

Figure 11.2 shows the number of publications in different categories where the X-axis represents categories and the Y-axis represents the number of publications. It displays that the quantity of articles is more than all categories, with 2,481 documents, which is 37% of total documents, followed by conference papers and book chapters, with 2,207 and 991 occupying 33% and 15%, respectively. When combining these three, they captured 85% of the total documents. At the same time, the remaining three sources—reviews, conference reviews, and books—have fewer publications and occupy 15% of the total. Using this analysis, the authors suggested that future researchers can conduct more review-based studies, whether they are reviews or conference reviews. This study also suggested writing of books and book chapters as students highly refer to them.

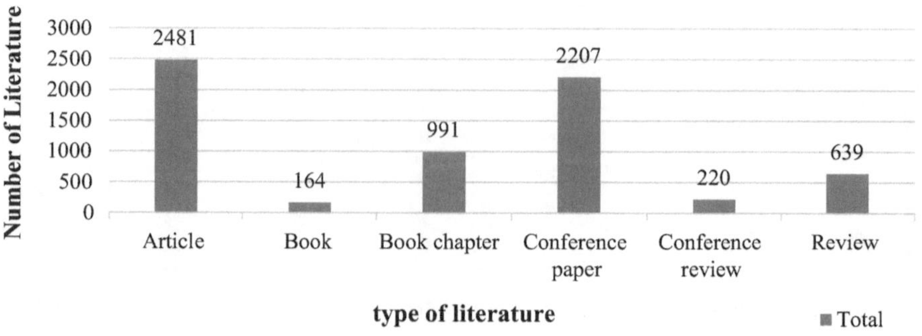


**FIGURE 11.2** Big data documents category analysis

### 11.2.2 Blockchain

Blockchain is an emerging topic that needs a standard definition. However, several authors expressed their own opinions related to this technology. According to Zīle and Strazdiņa, a blockchain is a chain of different cryptographic networks (blocks) that share a transaction with all the available nodes (users) in a network [47]. Similarly, a study also defined it as a technology of peer-to-peer networks that maintain a recorded transaction across several distributed nodes to provide rigidity and integrity for the data [43]. The main aim of blockchain is to provide tamper-proof storage for a valid transaction, designed so that any changes in existing blocks will be visible to all participants in a blockchain [36].

The process of blockchain starts with an agent by creating a transaction with a pre-defined rule and transmitting it for verification and validation purposes; once acceptance is received from the majority of the nodes, then the transaction will be added as a new block in a chain and saved in multiple nodes [2]. Besides, it provides a new way to store information, execute transactions, establish security, build public trust, and generate massive information exchange in a server [32]. Furthermore, this technology could disrupt significant industries such as real estate, healthcare, financial services, and more [38].

Blockchain technology has multiple benefits like business process optimization, involves low cost, provides security & privacy, interoperability, and enhances outcomes, which attracted the attention of healthcare industry. So, this industry is using blockchain technology for different purposes, such as patient information storage & tracking, transferring funds & supply chain management, revenue management, and securing clinical trials information [3]. Last year at the end of October, one of the biggest security breach was happened in India from the COVID-19 test details of Indian Council for Medical Research website. In this incident, over 81.5 crores of Indian citizens' information was stolen by hackers [16]. This type of incidents might get solution with the help of blockchain technology.

#### 11.2.2.1 Annual Scientific Production of Blockchain

Figure 11.3 of annual scientific production of blockchain ranged from 2016 to 2024 and had a total of 5,452 literature. Based on the previous discussion about 2024

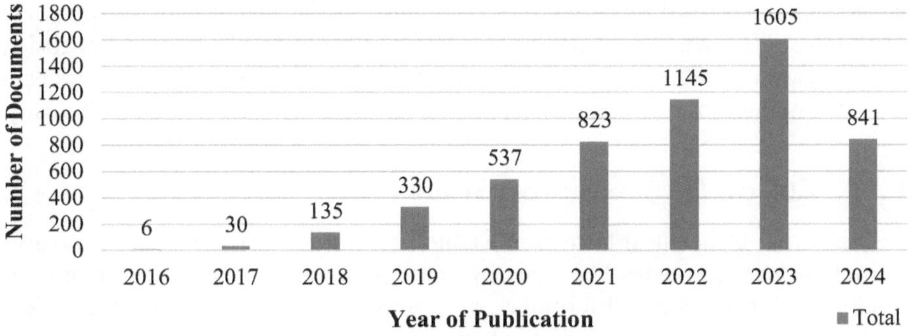


**FIGURE 11.3** Annual scientific production of blockchain

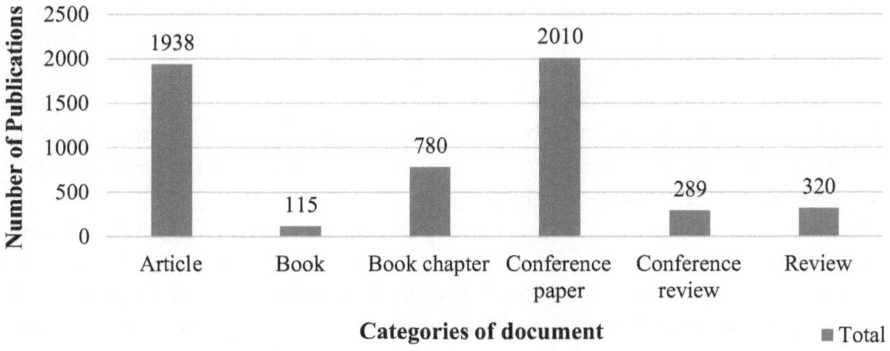


**FIGURE 11.4** Blockchain documents category analysis

publication in big data analysis, this study found that current available literatures are 52% of 2023 publications after completion of 60% of the current year (7/12 × 100). It shows that the current year publications in blockchain have possibility to grow more than 2023 publications. At the same time, the bar graph shows continuous rapid increase in number of publications in healthcare related to blockchain. The growth level is in such a high that the total publications raised to 140% from 2022 to 2023, but when comparing 2021 and 2023 the publications almost doubled. This finding shows the importance of blockchain in healthcare so the authors of this research suggested to future experts that the use of blockchain technology in healthcare will be benefitable.

#### 11.2.2.2 Blockchain Document Category Analysis

Concerning Table 11.1 and Figure 11.4, out of 5,452 documents published in blockchain, 2,010 were published in conference papers, followed by articles with 1,938 documents, which shows that researchers have a more significant target for conference papers and articles in terms of blockchain in healthcare. In general, only the conference papers and articles together captured 72% of total publications. Book chapters occupied third place with 780 literature. Even though book chapters are comparatively fewer than articles, when combined, the remaining three categories, such as reviews, conference reviews, and books, are still fewer than the book chapters. This finding shows an opportunity for future researchers to explore theoretically.

### 11.2.3 Clinical Decision Support System

CDSS is a tool of health information technology that is intended to integrate artificial intelligence in medical practice by developing and implementing algorithms and clinical information and formulating advanced diagnostic evaluation, forecasting of risk, prognostications, and treatment strategies to enhance better clinical outcomes [27]. The new methods of artificial intelligence with vast amounts of clinical data changed CDSS from a rule-based approach to a numeric-based approach [44] (Figure 11.5).

CDSS consists of four phases. The first stage deals with data collection related to patients and disease. Meanwhile, the second phase is divided into two sub-phases, namely, the inference phase and the risk aggression phase; again, this inference phase is classified into symbolic inference and statistical inference. The third phase determines corrected global risk based on the Breast Imaging Reporting and Data System indicator. Finally, the fourth phase is the final phase, which deals with providing warnings and making decisions [6].

The primary aim of the CDSS is to handle chronic conditions, providing alerts for medication prescriptions, notifications for immunization and preventive services, diagnoses, and treatment recommendations [19]. It also helps to reduce human errors and provides evidence-based clinical decision-making for pharmacists, doctors, and other professionals in healthcare [9]. However, the artificial intelligence–driven CDSS is still in the developing stage and much less in the clinical field, so it is unclear to forecast how these systems are handled by users [44].

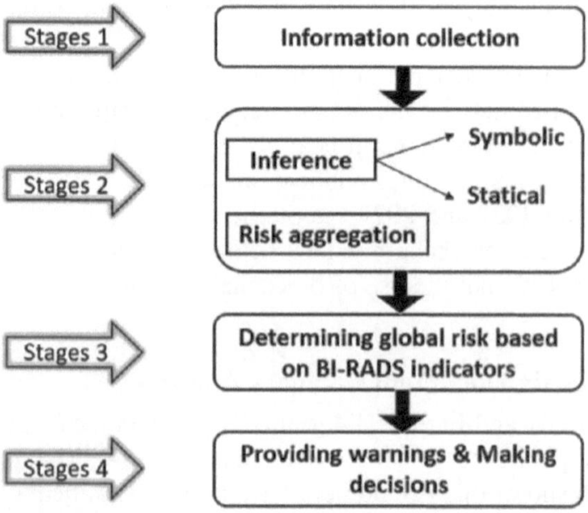


**FIGURE 11.5** Stages of clinical decision support system

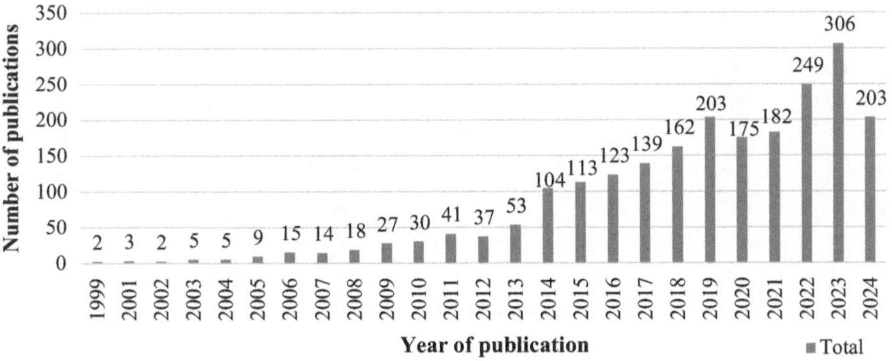


**FIGURE 11.6** Annual scientific production of clinical decision support system

#### 11.2.3.1 Annual Scientific Production of Clinical Decision Support System

Figure 11.6 shows that the publications in 2024 on CDSSs have more publications than all years except 2019, 2022, and 2023. Even though still five months are there for the completion of the year. It shows that researchers are more focused on CDSSs. Other than this year, when observing the remaining years of publications from 1999 to 2023, it is observed that there is no rapid growth in this field. However, it has been growing continuously for 24 years, which explains the stability of this area. Regarding the findings, the present study's authors encourage future researchers to continue their research in this field.

#### 11.2.3.2 CDSS Document Category Analysis

After filtration in healthcare, the total number of available documents in the CDSS was 2,220, of which 1,237 are articles, meaning 56% of the total documents in CDSS. Next, conference papers and reviews captured the second and third positions, with 474 and 367. Observing previous big data and blockchain analyses, articles, conference papers, and book chapters always occupied the top positions. However, with the current technology, reviews have gained a third place, more than three times the number of book chapters. Finally, the last two places were the same in all of these three analyses; it might be due to the preference of researchers to perform conference papers rather than conference reviews. In the case of books, quantity and time might affect publications. For a better understanding, Figure 11.7 is shown below.

### 11.2.4 Internet of Things

The term IoT was first tossed by computer scientist "Kevin Ashton" in 1999 [10]. It is a network of multiple electronic devices that can communicate with each other through a communication protocol and have the capacity to process the data [46]. Similarly, a study says that the IoT is the process of information sharing among multiple devices, buildings, vehicles, and other things through sensors, software, or a network to enhance efficiency and productivity [21].

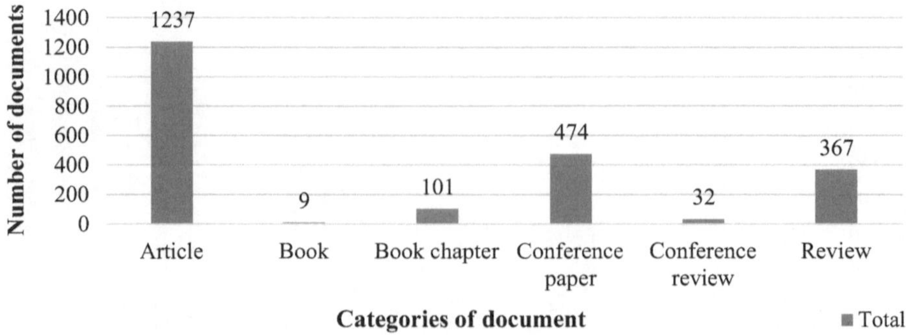


**FIGURE 11.7** Clinical decision support system documents category analysis

Generally, IoT consists of three layers of architecture: perception, network, and application. The perception layer is the primary and physical layer that identifies and gathers information with the help of sensors. However, the network layer is the brain of IoT, which acts as a hub for network management, internet networks, and intelligent processing. Finally, the application layer, also known as the business layer, is located at the top of the structure and provides application services by providing authentication and maintaining confidentiality and data integrity [20].

The number of IoT devices worldwide will be doubled from 15.1 billion in 2020 to more than 29 billion by 2030 [42]. Consequently, this rapid growth creates a vast amount of data, which will become challenging for networks to handle efficiently [21]. On the other hand, it has a great advantage in medical field by enhancing healthcare with real-time remote monitoring, robotic surgeries, smart inhalers, and more [10], which results in frequently increasing the spending on IoT in healthcare sector.

### 11.2.5 Annual Scientific Production of Internet of Things

Based on observation of the annual scientific production of IoT in healthcare, it was found that the publications in healthcare started in 2006 with 1 document and rose to 3,113 after 18 years in 2023. Figure 11.8 shows current- and future-year literature because of the consideration of published and ready-to-publish (article in press) documents. As the study is related to the current year, the publication in 2025 is negligible. Now, when looking into the publications of 2024, 1,534 documents are currently available, which is nearly 50% of published documents in 2023; thus, the authors understand that this area has a scope to increase the total number of publications to previous years. Additionally, the study found that there is a rapid growth in IoT. With this, the study suggests to future researchers that the concept of IoT is old. However, it is still trending, so experts can consider IoT for research.

### 11.2.6 IoT Document Category Analysis

Figure 11.9 of the category of document analysis related to IoT in healthcare reflects a similar pattern to blockchain category analysis. The total number of publications in

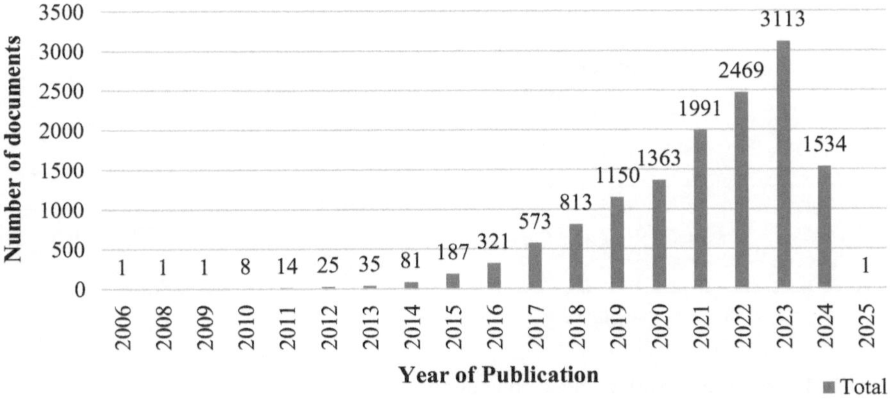


**FIGURE 11.8** Annual scientific production of internet of things

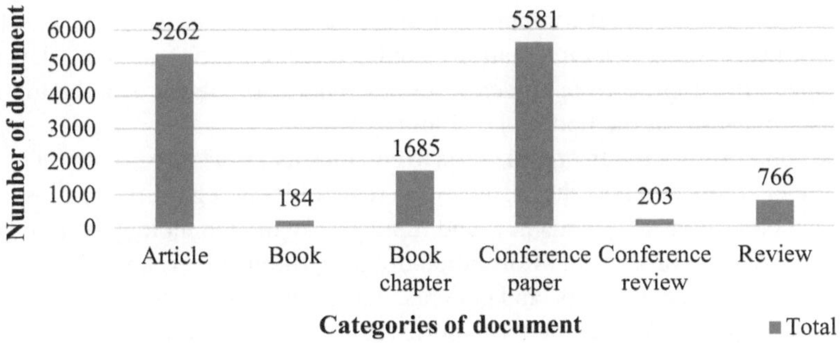


**FIGURE 11.9** Internet of things documents category analysis

IoT is 13,681, of which both conference papers and articles have 79% of documents. This indicates that researchers are highly focused on empirical rather than theoretical research. Based on this finding, future researchers are supposed to split their focus into theoretical research to enhance current knowledge and provide more opportunities for future empirical research.

## 11.3 CLOUD COMPUTING

Eric Schmidt, an executive officer of Google, first tossed the term cloud computing in 2006. Later, it was treated as a prominent part of Industrial 4.0 [1]. Cloud computing is a subscription-based software service provided by the vendor with a license at a low cost and high security, and it also provides necessary hardware devices on a rental basis upon request. It gives the organization complete control over security and private IT space. In simple terms, cloud computing is a service provided to the client by the vendors as a server to store information through the internet [25].

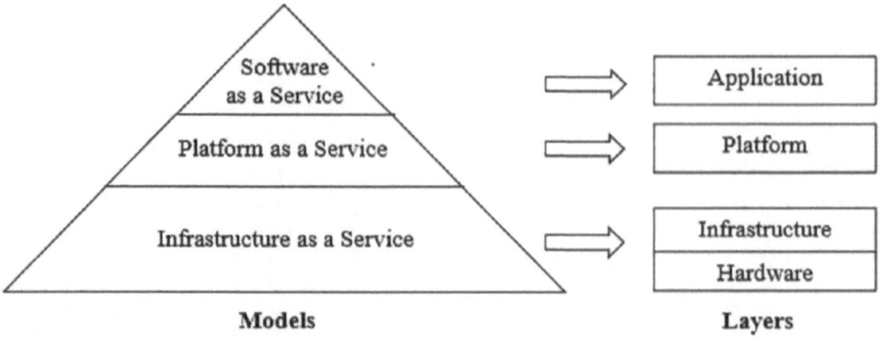


**FIGURE 11.10** Framework of cloud computing

It consists of three models, which include Software as a Service (SaaS), Platform as a Service (PaaS), and Infrastructure as a Service (IaaS), to provide services to users. Furthermore, IaaS is at the lower core, PaaS is at the middle core, and SaaS is at the upper core of cloud computing. These models are classified into four layers such as hardware, infrastructure, platform, and application. However, these layers are also arranged in a top-down approach where the application layer is at the top while the hardware layer is at the bottom [1] (Figure 11.10).

Cloud computing helps an organization to enhance its IT resources without upgrading equipment, which makes the organization react quickly regarding business needs and market conditions [17]. This new and rapidly growing technology is used [12] in healthcare industry for sharing information in a structured and organized way among health professionals, caretakers, and patients, which reduces response time, enhances service quality, and provides better experience to the patients [8]. Despite various advantages, it holds a great threat from the external partners of cloud computing in healthcare where maintaining confidentiality of patient information is mandatory [12]. Along with this, the users and service providers may experience dissatisfaction when distributed denial of services or high response time of services [15].

### 11.3.1 Annual Scientific Production of Cloud Computing

After CDSSs and IoT, cloud computing is covered for an extended period from 2009 to 2025. Based on previous discussions, publications of 2025 were not considered. Figure 11.11 shows a similar trend of previous technologies, and the publications in 2024 also show more than 50% of publications related to 2023, so in simple terms, the exact directions of previous research are applicable here.

### 11.3.2 Cloud Computing Document Category Analysis

By looking for the similarity pattern, the pattern shown in Figure 11.12 is aligned with the big data pattern that will provide similar future directions. However, as per the discussion, cloud computing has 4,786 documents and is fourth in the number of documents after IoT, big data, and blockchain. Of 4,786 documents, 79% are

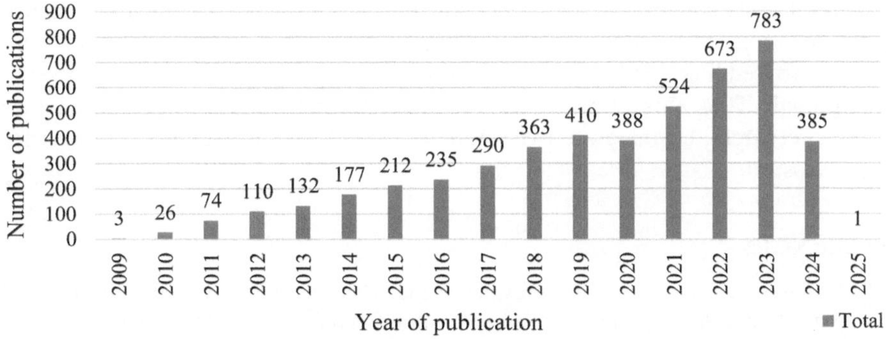


**FIGURE 11.11** Annual scientific production of cloud computing

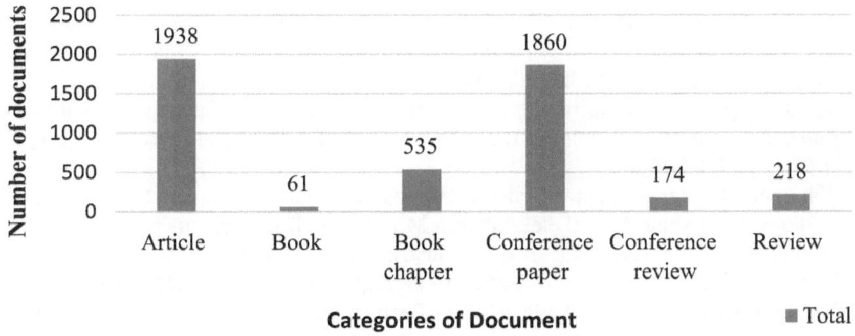


**FIGURE 11.12** Cloud computing documents category analysis

articles and conference papers. After these, third place was gained by book chapters, followed by reviews, conference reviews, and books.

## 11.4 CONCLUSION AND FUTURE DIRECTIONS

The rapidly changing technology is continuously making day-to-day tasks more manageable. In recent years, some disruptive technologies have come into existence to replace the traditional ways of doing things and provide alternative ways to accomplish them. The current study considered five technologies: big data, blockchain, CDSS, IoT, and cloud computing, used in healthcare to provide customized services for patients and make hospitals from hospital-centric to patient-centric. This research provides a detailed explanation of each technology's benefits and challenges. Additionally, this study performed annual production analysis and category of document analysis for all five technologies.

This analysis concluded that most researchers are focused on empirical research, like articles and conference papers, but give less importance to theoretical research. This study suggests that future experts perform more review-based studies to enhance the current understanding of these technologies and provide theoretical assistance for future empirical research. The annual scientific production shows that all technologies are trending in nature, with some showing rapid growth and some showing slight growth. Thus, this study suggests future researchers to use these technologies for their research and integrate two or more technologies to identify solutions for complex issues.

## REFERENCES

1. Abdullayeva, Fargana J. "Cyber resilience and cyber security issues of intelligent cloud computing systems." *Results in Control and Optimization* 12 (September 1, 2023): 100268. https://doi.org/10.1016/j.rico.2023.100268.
2. Akhtar, Pervez, Nora Azima, Abdul Ghafar, and Shahab Ud Din. "Barricades in the adoption of block-chain technology in supply chain management: Challenges and benefits." *Transnational Marketing Journal* 9, no. 1 (May 1, 2021): 3–16. https://doi.org/10.33182/tmj.v9i1.1021.
3. Attaran, Mohsen. "Blockchain technology in healthcare: Challenges and opportunities." *International Journal of Healthcare Management* 15, no. 1 (November 8, 2020): 70–83. https://doi.org/10.1080/20479700.2020.1843887.
4. Bellucci, Nancy. "Disruptive innovation and technological influences on healthcare." *Journal of Radiology Nursing* 41, no. 2 (June 1, 2022): 98–101. https://doi.org/10.1016/j.jradnu.2022.02.008.
5. Bongomin, Ocident, Aregawi Yemane, Brendah Kembabazi, Clement Malanda, Mwewa Chikonkolo Mwape, Nonsikelelo Sheron Mpofu, and Dan Tigalana. "Industry 4.0 disruption and its neologisms in major industrial sectors: A state of the art." *Journal of Engineering* 2020 (October 10, 2020): 1–45. https://doi.org/10.1155/2020/8090521.
6. Casal-Guisande, Manuel, Antía Álvarez-Pazó, Jorge Cerqueiro-Pequeño, José-Benito Bouza-Rodríguez, Gustavo Peláez-Lourido, and Alberto Comesaña-Campos. "Proposal and definition of an intelligent clinical decision support system applied to the screening and early diagnosis of breast cancer." *Cancers* 15, no. 6 (March 10, 2023): 1711. https://doi.org/10.3390/cancers15061711.
7. Coherent Market Insights. "Healthcare cloud computing market to surpass US 1738863 Mn by 2030- Coherent market insights." *Coherent Market Insights*, October 14, 2022. Accessed November 29, 2023. https://www.coherentmarketinsights.com/press-release/cloud-computing-market-in-healthcare-industry-3413.
8. Dang, L. Minh, Md. Jalil Piran, Dongil Han, Kyungbok Min, and Hyeonjoon Moon. "A survey on internet of things and cloud computing for healthcare." *Electronics* 8, no. 7 (July 9, 2019): 768. https://doi.org/10.3390/electronics8070768.
9. Detopoulou, Paraskevi, Πάνος Παπανδρέου, L. Papadopoulou, and Maria Skouroliakou. "Implementation of a Nutrition-Oriented Clinical Decision Support System (CDSS) for weight loss during the COVID-19 epidemic in a hospital outpatient clinic: A 3-month controlled intervention study." *Applied Sciences* 13, no. 16 (August 21, 2023): 9448. https://doi.org/10.3390/app13169448.
10. Dlewis. "IoT technologies explained: History, examples, risks & future." *Vision of Humanity*, April 22, 2022. Accessed January 19, 2024. https://www.visionofhumanity.org/what-is-the-internet-of-things/.
11. Franceschi, Matteo, Davide Morelli, David Plans, Alan W. Brown, John Collomosse, Louise V. Coutts, and Laura Ricci. *ComeHere: Exploiting Ethereum for Secure Sharing of Health-Care Data.* Springer eBooks, 585–596, 2018. https://doi.org/10.1007/978-3-030-10549-5_46.

12. Griebel, Lena, Hans-Ulrich Prokosch, Felix Köpcke, Dennis Toddenroth, Jan Christoph, Ines Leb, Igor Engel, and Martin Sedlmayr. "A scoping review of cloud computing in healthcare." *BMC Medical Informatics and Decision Making* 15, no. 1 (March 19, 2015): 1–16. https://doi.org/10.1186/s12911-015-0145-7.
13. Gupta, Aditya, and Amritpal Singh. "Healthcare 4.0: Recent advancements and futuristic research directions." *Wireless Personal Communications* 129, no. 2 (December 28, 2022): 933–952. https://doi.org/10.1007/s11277-022-10164-8.
14. Hassani, Hossein, and Emmanuel Sirimal Silva. "Big data: A big opportunity for the petroleum and petrochemical industry." *OPEC Energy Review* 42, no. 1 (March 1, 2018): 74–89. https://doi.org/10.1111/opec.12118.
15. Hezavehi, Sasha Mahdavi, and Rouhollah Rahmani. "Interactive anomaly-based DDoS attack detection method in cloud computing environments using a third party auditor." *Journal of Parallel and Distributed Computing* 178 (August 1, 2023): 82–99. https://doi.org/10.1016/j.jpdc.2023.04.003.
16. HT News Desk. "Aadhaar details of 81.5 cr people leaked in India's 'biggest' data breach." *Hindustan Times*, October 31, 2023. Accessed January 18, 2024. https://www.hindustantimes.com/technology/in-indias-biggest-data-breach-personal-information-of-81-5-crore-people-leaked-101698719306335.html.
17. Islam, Rafia, Vardhan Patamsetti, Aparna Gadhi, Ragha Madhavi Gondu, Chinna Manikanta Bandaru, Sai Chaitanya Kesani, and Olatunde Abiona. "The future of cloud computing: benefits and challenges." *International Journal of Communications, Network and System Sciences* 16, no. 04 (January 1, 2023): 53–65. https://doi.org/10.4236/ijcns.2023.164004.
18. Jahani, Hamed, Richa Jain, and Dmitry Ivanov. "Data science and big data analytics: A systematic review of methodologies used in the supply chain and logistics research." *Annals of Operations Research*, July 11, 2023. https://doi.org/10.1007/s10479-023-05390-7.
19. Jing, Xia, Hua Min, Yang Gong, Paul G. Biondich, David Robinson, Timothy Law, Christian Nøhr, et al. "Ontologies applied in clinical decision support system rules: Systematic review." *JMIR Medical Informatics* 11 (January 19, 2023): e43053. https://doi.org/10.2196/43053.
20. Kakkar, L., Gupta, D., Saxena, S., & Tanwar, S. (2021). IoT architectures and its security: A review. In *Proceedings of the Second International Conference on Information Management and Machine Intelligence: ICIMMI 2020* (pp. 87–94). Singapore: Springer.
21. Kamarudin, Imran Edzereiq, Mohamed Ariff Ameedeen, Mohd Faizal Ab Razak, and Azlee Zabidi. "Software defined internet of things in smart city: A review." *Indonesian Journal of Electrical Engineering and Computer Science* 32, no. 2 (November 1, 2023): 915. https://doi.org/10.11591/ijeecs.v32.i2.pp915-924.
22. Khan, Nawsher, Ibrar Yaqoob, Mohamed Hashem, Zakira Inayat, Waleed Kamaleldin Mahmoud Ali, Muhammad Mansoor Alam, Muhammad Shiraz, and Abdullah Gani. "Big data: Survey, technologies, opportunities, and challenges." *The Scientific World Journal* 2014 (January 1, 2014): 1–18. https://doi.org/10.1155/2014/712826.
23. Khanra, Sayantan, Amandeep Dhir, A. K. M. Najmul Islam, and Matti Mäntymäki. "Big data analytics in healthcare: A systematic literature review." *Enterprise Information Systems* 14, no. 7 (August 8, 2020): 878–912. https://doi.org/10.1080/17517575.2020.1812005.
24. Li, Fangyu, Yuanjun Laili, X. Chen, Yihuai Lou, Chen Wang, Hongyan Yang, Xuejin Gao, and Hugang Han. "Towards big data driven construction industry." *Journal of Industrial Information Integration* 35 (October 1, 2023): 100483. https://doi.org/10.1016/j.jii.2023.100483.

25. Madhavaiah, C., Irfan Bashir, and Syed Irfan Shafi. "Defining cloud computing in business perspective: A review of research." *Vision* 16, no. 3 (September 1, 2012): 163–173. https://doi.org/10.1177/0972262912460153.
26. Meticulous Market Research Pvt. Ltd. "Internet of Things (IoT) in healthcare market is expected to grow at a CAGR of 29.9% to Reach $322.2 billion by 2025: Meticulous Research®." *GlobeNewswire News Room*, March 19, 2020. Accessed November 28, 2023. https://www.globenewswire.comnews-release/2020/03/19/2003195/0/en/Internet-of-Things-IoT-in-Healthcare-Market-is-Expected-to-Grow-at-a-CAGR-of-29-9-to-Reach-322-2-billion-by-2025-Meticulous-Research.html.
27. Misro, Aswini, Anushka Mehta, Paul Whittington, Hüseyin Doğan, Nishikant Mishra, Naim Kadoglou, and Selva Theivacumar. "Analysis of innovation adoption challenges and opportunities surrounding Clinical Decision Support Systems (CDSS) in India." *medRxiv* (Cold Spring Harbor Laboratory), April 3, 2023. https://doi.org/10.1101/2023.04.02.23288046.
28. Oussous, Ahmed, Fatima Zahra Benjelloun, Ayoub Ait Lahcen, and Samir Belfkih. "Big Data technologies: A survey." *Journal of King Saud University - Computer and Information Sciences* 30, no. 4 (October 1, 2018): 431–48. https://doi.org/10.1016/j.jksuci.2017.06.001.
29. Prakash, C., R.Yadav, and S. Kadyan. "Effect of the price drop on customer's perceived evaluation across selected product categories." *Journal of Revenue and Pricing Management* 20 (2021): 204–210.
30. Precedence research. "Clinical decision support systems market size, trends, report 2032." August 2023. Accessed November 29, 2023. https://www.precedenceresearch.com/clinical-decision-support-systems-market#:~:text=The%20global%20clinical%20decision%20support,8.3%25%20from%202023%20to%202032.&text=As%20there%20is%20an%20increasing,and%20integrated%20reliable%20technical%20solutions.
31. Rahul Rahul, Kumar, Rohitash Kumar Banyal, and Neeraj Arora. "A systematic review on big data applications and scope for industrial processing and healthcare sectors." *Journal of Big Data* 10, no. 1 (August 27, 2023): 133. https://doi.org/10.1186/s40537-023-00808-2.
32. RawalBharat, S., M. Poongodi, ManogaranGunasekaran, and HamdiMounir. "Multi-Tier stack of block chain with proxy re-encryption method scheme on the internet of things platform." *ACM Transactions on Internet Technology* 22, no. 2 (October 29, 2021): 1–20. https://doi.org/10.1145/3421508.
33. Ristevski, Blagoj, and Ming Chen. "Big data analytics in medicine and healthcare." *Journal of Integrative Bioinformatics* 15, no. 3 (May 10, 2018): 1–24. https://doi.org/10.1515/jib-2017-0030.
34. Sa, Senthilkumar, K Bharatendara, Amruta A Meshram, Angappa Gunasekaran, and S. Chandrakumarmangalam. "Big data in healthcare management: A review of literature." *American Journal of Theoretical and Applied Business* 4, no. 2 (January 1, 2018): 57. https://doi.org/10.11648/j.ajtab.20180402.14.
35. Sagıroğlu, Şeref, and Duygu Sinanç. "Big data: A review." *International Conference on Collaboration Technologies and Systems (CTS)*, May 1, 2013. https://doi.org/10.1109/cts.2013.6567202.
36. Samaniego, Mayra, and Ralph Deters. "Using blockchain to push software-defined IoT components onto edge hosts." In *Proceedings of the International Conference on Big Data and Advanced Wireless Technologies*, November 10, 2016. https://doi.org/10.1145/3010089.3016027.
37. Samhan, Bahae, and Kailash Joshi. "Disruptive-technology avoidance in healthcare." *International Journal of Healthcare Information Systems and Informatics* 14, no. 2 (April 1, 2019): 28–48. https://doi.org/10.4018/ijhisi.2019040103.

38. Sarwan, Kartik, Sanket Thore, Niranjan Satav, Pruthviraj Kamble, and Dibyo Mandal. "A secure e-wallet system using block chain." *International Journal of Scientific Research in Computer Science, Engineering and Information Technology* 8, no. 3 (May 25, 2021): 454–466. https://doi.org/10.32628/cseit2173101.
39. Schwamm, Lee H. "Telehealth: Seven strategies to successfully implement disruptive technology and transform health care." *Health Affairs* 33, no. 2 (February 1, 2014): 200–206. https://doi.org/10.1377/hlthaff.2013.1021.
40. Statista, and Stewart. "Global big data healthcare analytics market size by application 2016 & 2025." *Statista*, October 9, 2021. Accessed November 28, 2023. https://www.statista.com/statistics/909669/global-big-data-in-healthcare-analytics-market-size-by-application/.
41. Tortorella, Guilherme Luz, Flávio Sanson Fogliatto, Alejandro Vergara, Roberto Vassolo, and Rapinder Sawhney. "Healthcare 4.0: Trends, challenges and research directions." *Production Planning & Control* 31, no. 15 (December 17, 2019): 1245–1260. https://doi.org/10.1080/09537287.2019.1702226.
42. Vailshery, Lionel Sujay. "IoT connected devices worldwide 2019–2030 | Statista." *Statista*, July 27, 2023. Accessed January 1, 2024. https://www.statista.com/statistics/1183457/iot-connected-devices-worldwide/.
43. Viriyasitavat, Wattana, and Danupol Hoonsopon. "Blockchain characteristics and consensus in modern business processes." *Journal of Industrial Information Integration* 13 (March 1, 2019): 32–39. https://doi.org/10.1016/j.jii.2018.07.004.
44. Wang, Dakuo, Liuping Wang, Zhan Zhang, Ding Wang, Haiyi Zhu, Yvonne Y. Gao, Xiangmin Fan, and Feng Tian. "'Brilliant AI Doctor' in rural clinics: Challenges in AI-powered clinical decision support system deployment." *Proceedings of the 2021 CHI Conference on Human Factors in Computing Systems*, May 6, 2021. https://doi.org/10.1145/3411764.3445432.
45. Yaqoob, Ibrar, Mohamed Hashem, Abdullah Gani, Salimah Binti Mokhtar, Ejaz Ahmed, Nor Badrul Anuar, and Athanasios V. Vasilakos. "Big data: From beginning to future." *International Journal of Information Management* 36, no. 6 (December 1, 2016): 1231–1247. https://doi.org/10.1016/j.ijinfomgt.2016.07.009.
46. Yıldırım, Muhammed, Uğur Demiroğlu, and Bilal Şenol. "An in-depth exam of IoT, IoT core components, IoT layers, and attack types." *Europan Journal of Science and Technology*, no. 14 (October 16, 2021). https://doi.org/10.31590/ejosat.1010023.
47. Zīle, Kaspars, and Renāte Strazdiņa. "Blockchain use cases and their feasibility." *Applied Computer Systems* 23, no. 1 (May 1, 2018): 12–20. https://doi.org/10.2478/acss-2018-0002.

# 12 Embracing Technological Advancements

## *The Transformation of Healthcare Services in the Digital Age*

*Kunal Rawal, Anushree Acharya, and Shweta Soni*

## 12.1 INTRODUCTION

Digital health, an expansive term encapsulating the convergence of technology and healthcare, is defined comprehensively by the World Health Organization (WHO) [1]. It encompasses e-health alongside burgeoning fields like big data analytics, genomics, and artificial intelligence. Technology has emerged as a pivotal cornerstone in healthcare, propelling significant advancements.

In recent years, a notable surge has been witnessed across various dimensions of digital health, spanning wearable technology, telemedicine, online pharmacy services, and other innovative products [2]. Moreover, substantial research and development endeavors have been directed towards integrating cutting-edge technologies such as blockchain, virtual reality, robotics, and artificial intelligence with pharmaceuticals and healthcare. Notably, the chasm between effective healthcare delivery via technology and traditional methods is gradually diminishing, mainly due to ambient computing approaches.

Recognizing the transformative potential of digital health, the Government of India initiated the flagship Digital India Campaign in 2015. This initiative aimed to leverage digital technologies to bolster healthcare access, particularly in rural areas. Subsequently in 2018, the Digital Health Mission in India was formally inaugurated, aligning with the vision delineated in the National Health Policy (NHP) 2017 for a fully digitalized healthcare system. India's steadfast commitment to digital health, buoyed by governmental attention and resources, has positioned the nation as a conducive market for healthcare innovation and investment opportunities [3]. The impetus for India's venture into digital public health stemmed from significant population growth, leading to a notable disparity between healthcare providers and patients, escalating healthcare costs. Digital health interventions have showcased promising outcomes, encompassing cessation programs, maternal health education, addressing


DOI: 10.1201/9781032700953-14

mental health challenges, and optimizing vaccine supply chains. Consequently, digitalizing public health has emerged as a pressing necessity in India.

India's strides in the digital health realm have garnered international acclaim, epitomized by India's leadership in spearheading a landmark resolution on digital health for health system strengthening at the 71st World Health Assembly under the auspices of the WHO. This commitment culminated in launching the National Digital Health Mission (NDHM), a pioneering initiative unveiled by the honorable Prime Minister of India on August 15, 2020. India's trajectory towards the NDHM was shaped by earlier initiatives such as the NHP and the National Digital Health Blueprint (NDHB).

Despite advancements across various sectors, the healthcare industry must still catch up and embrace digitalization. One of the primary impediments to progress lies in concerns expressed by medical professionals, patients, and other stakeholders regarding the security and accessibility of their digital medical records. However, the transformative impact of digital advancements in healthcare cannot be overstated, offering individuals enhanced accessibility to healthcare services, even in remote or underserved regions. This paradigm shift reshapes healthcare delivery conceptually and operationally [4].

## 12.2 NEED FOR DIGITALIZATION OF HEALTHCARE IN INDIA

India's healthcare sector is undergoing a profound digital transformation spurred by the widespread adoption of smartphones and the increasing accessibility of internet connectivity. Each year witnesses a remarkable 30% growth in smartphone usage and internet access nationwide. This digital revolution owes much of its momentum to the vibrant Indian startup ecosystem, significantly contributing to modernizing the healthcare industry.

To fully capitalize on digitalization, businesses actively embrace the SMAC (Social, Mobile, Analytics, Cloud) architecture and foster a digital mindset. This strategic pivot aims to unearth new avenues for value creation within the healthcare landscape [5]. Public institutions and multinational corporations fervently advocate for integrating digital technologies into India's healthcare sector.

The term "digital strategy" encompasses a comprehensive approach to disrupting traditional corporate operations, ensuring enhanced and sustained access to healthcare services for all. It is crucial to recognize that technology does not offer a one-size-fits-all solution; instead, it empowers the development of innovative healthcare solutions, enhances patient care, and expands provider networks [6].

Given the pivotal role of the healthcare sector in safeguarding human health and well-being, a continuous process of upgrade and evolution is imperative. Enhancing the healthcare system, in turn, augments its overall efficacy and efficiency. The global impact of e-health has been transformative, fostering a more informed and open-minded society [7]. One of the most promising aspects of this digital revolution is its potential to bridge the existing gap between patients and healthcare providers. By transcending the constraints of time and space, digital healthcare promises to make services more accessible to all, ultimately advancing healthcare accessibility on a global scale [8].

Embarking on this challenging new global mission necessitates the healthcare system to reinvent and reevaluate its infrastructure. This evolution aims to liberate care delivery, aligning with patients' evolving expectations and public policy criteria while addressing ever-evolving technical requirements and goals [9].

There is an unmistakable demand for reliable and cost-effective healthcare services in India. The government's commitment to increasing healthcare spending to 2.5% of GDP in the 12th five-year plan from 2022, with a future goal of reaching 3%, underscores the pressing need for digital transformation within the Indian healthcare sector. This commitment reflects the rising importance of leveraging digital technology to meet the nation's evolving healthcare needs (Figure 12.1).

i. **Healthcare service optimization**: The main objective of digitalization is to deliver fast, precise, and efficient health services.
ii. **Better diagnosis**: The system offers enhanced processing skills that could not be achieved with the old system, medical imaging like X-rays and cardiographs.
iii. **Ensured integration**: Exchanging patient data helps make obvious diagnoses and increases the reliability of corrective and preventative responses to new data.
iv. **Guaranteed openness and support**: Versatility; where can I locate the most current, offline, and online health records? What will I find? The issue is resolved because data that spans space and time is accessible around the clock.
v. **Integrity and safety**: To offer open and secure facilities for utilizing resources and care.

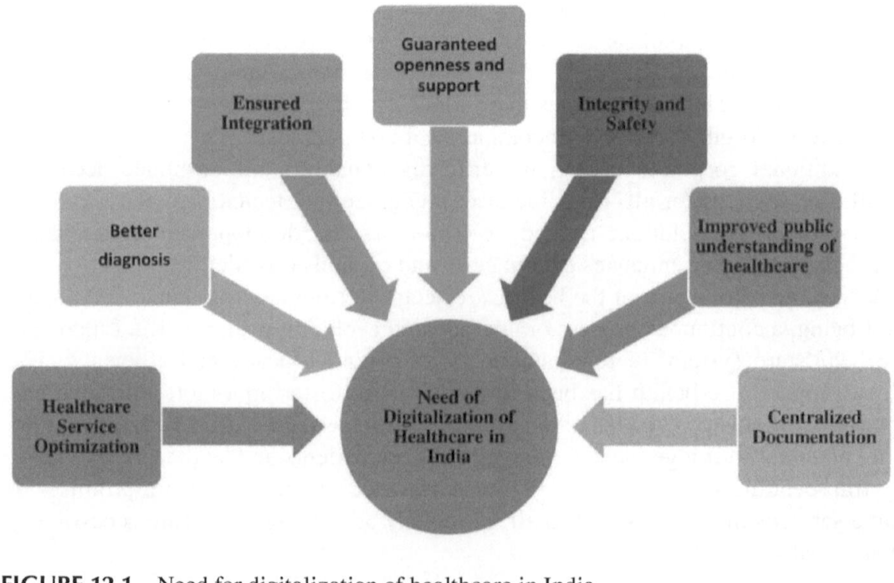


**FIGURE 12.1** Need for digitalization of healthcare in India

vi. **Improved public understanding of healthcare**: A program that promotes knowledge of medical resources, medications, available doctors, ratings of services provided, patient legal rights, and other information based on actual requirements has been implemented.
vii. **Centralized documentation**: Because patient information is frequently disseminated over numerous networks and locations via various techniques, centralized access to patient records is generally impossible. A successful medical involvement might begin with a safe and easily accessible medical history.

## 12.3 THE INCLUSION OF DIGITALIZATION IN THE SUSTAINABLE DEVELOPMENT AGENDA 2030

The 2030 Agenda for Sustainable Development emphasizes the significant potential for advancing human development, narrowing the digital divide, and fostering knowledge-based societies. This vision materializes through the widespread adoption of information and communications technology and global connectivity [10]. Notable technological advancements were celebrated during a high-level gathering convened by the UN General Assembly on December 15–16, 2015, to assess the outcomes of the World Summit on the Information Society. These advancements have revolutionized the landscape of government healthcare delivery, extending access to services and information that were previously inaccessible or unaffordable for many individuals [11].

At this event, leaders of delegations and participating officials made solemn commitments to fully harness data and communication technology to pursue Sustainable Development Goals related to healthcare. They underscored the potential for technology to expedite progress in this critical domain, recognizing its pivotal role in advancing development objectives [12] (Figure 12.2).

## 12.4 DIGITALIZATION PROCESS IN THE HEALTHCARE SECTOR

The ongoing digitalization journey within the healthcare industry necessitates extensive consultation with stakeholders deeply vested in this transformative process. Elevating the standard of healthcare service delivery in the digital era demands more than mere modifications to existing components; it calls for a comprehensive reimagining of service processes [13]. It is crucial to acknowledge that the landscape of technology is in a perpetual state of evolution, marked by constant advancements. Thus, ensuring a successful digitalization endeavor requires agility and adaptability, with a commitment to integrating emerging technologies into the healthcare ecosystem as they mature.

A crucial precursor to technology deployment in healthcare is rigorous testing to validate its efficacy in real-world scenarios. Only after comprehensive evaluation can technology be deemed suitable for addressing the needs and desires of the populace. Should it fall short, enhancements must be undertaken to align with user demands [14].

**FIGURE 12.2** New Sustainable Development Goals transformation in digital healthcare

Two scenarios emerge when considering technology adoption within the healthcare sector: successful adoption and iterative improvement. In the former, technology seamlessly aligns with healthcare needs, enhances service delivery, and meets user expectations, making it readily incorporable into the healthcare ecosystem. Conversely, refinement and optimization are imperative to ensure alignment with healthcare requirements if technology initially falls short.

In navigating the intricate landscape of healthcare digitalization, flexibility, adaptability, and a relentless focus on enhancing the quality and effectiveness of healthcare services remain paramount. Collaboration among stakeholders and a commitment to iterative improvement are essential to realize the full potential of digitalization in healthcare [15] (Figure 12.3).

In the realm of technology adoption within the healthcare industry, two distinct scenarios often come into play:

a. **Technology Push**: In this scenario, technology providers initiate discussions with healthcare service providers regarding integrating their technologies into the healthcare ecosystem. This interaction might involve implementing

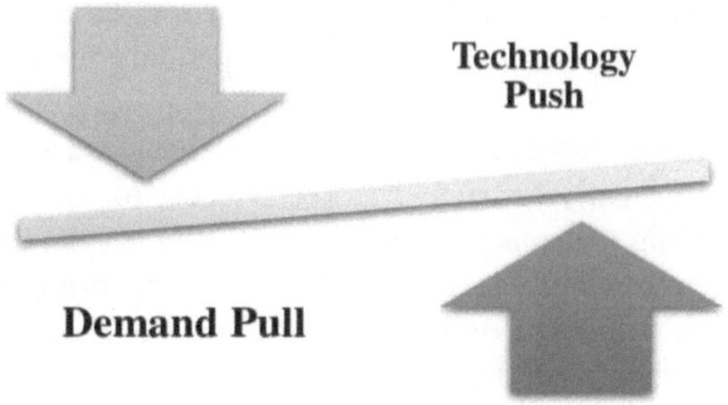


**FIGURE 12.3** Digitalization process in the healthcare sector

a pilot project or deploying technology on a larger scale. The healthcare service providers have yet to commit to purchasing the technology; they are still in the evaluation phase. Embracing digitalization can be a more challenging proposition when technology is introduced in this manner, as it requires convincing and demonstrating the value of the technology to potential users.

b. **Demand pull**: Contrasting with the technology-push approach, the demand-pull scenario arises when the healthcare industry identifies an existing issue or challenge. Healthcare service providers recognize a problem they are currently facing and actively seek out innovative technological solutions. They may choose to adapt existing technology to address the issue or collaborate with a specialized technology team to develop a tailored solution. Typically, this approach is more aligned with achieving specific objectives, as the healthcare teams have already identified the tools required for implementation.

In essence, these two scenarios offer different entry points for technology adoption in healthcare. In contrast, technology push involves proactive outreach from technology providers; demand-pull stems from the healthcare industry's inherent recognition of a need and the proactive pursuit of technology to meet that need. Both approaches have their merits and challenges, and successful digitalization efforts often hinge on effective communication and alignment between technology and healthcare service providers [16].

## 12.5 DIGITAL MODE OF HEALTHCARE SERVICES DURING THE PANDEMIC

The COVID-19 pandemic has unleashed a wave of challenges across various sectors, sparing none, particularly the economic and healthcare realms. Among its profound repercussions, the pandemic has significantly impacted the healthcare industry, prompting the rapid development of innovative and cutting-edge medical technologies.

As the pandemic has evolved, individuals worldwide have embraced many preventive medical measures and precautions, fostered heightened awareness, and prompted exploration of options in medical care, pharmaceuticals, and healthcare services. The exigencies of the pandemic have necessitated a creative overhaul of how the healthcare sector meets the unmet needs of patients during these challenging times.

A significant driver of these changes has been the government-imposed lockdowns, which have physically restricted people's access to healthcare facilities and establishments, including pharmacies. This has not only accelerated the adoption of digital healthcare solutions but has also shed light on the strengths, prospects, and areas for improvement within the digitalization of healthcare systems.

The pandemic has underscored the critical importance of three pivotal aspects of healthcare: access, cost, and quality. Remarkably, in the pre-COVID era, the concept of doctors diagnosing patients remotely or digitally was a distant possibility. However, it has swiftly become the prevailing practice in the current COVID-19 landscape. Consequently, India has redirected its focus towards comprehending how digital technologies reshape the healthcare sector, usher in new opportunities, and redefine how healthcare is delivered and accessed (Figure 12.4).

a. **M-Health**: With a projected market size of Rs. 2,083 crores, which would rise to Rs. 5,184 crores by 2020, it is one of the largest sectors in India's digital healthcare industry. M-Health is recommended by 68% of doctors, and around 59% of patients already use it. Mobile apps are vital to the M-Health

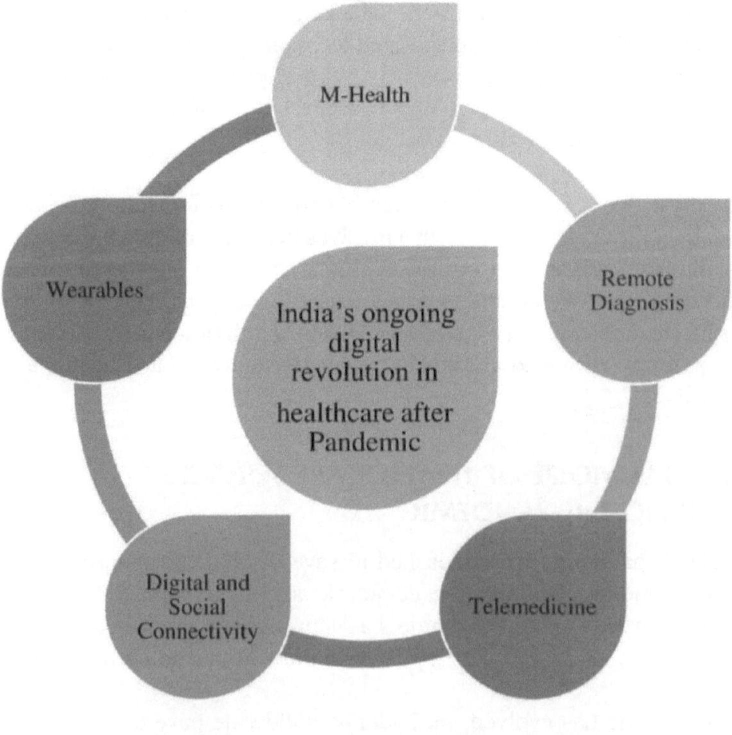


**FIGURE 12.4** India's ongoing digital revolution in healthcare after the pandemic

market because they typically connect patients and physicians and offer remote consultations.

b. **Remote Diagnosis**: Through the provision of point-of-care diagnostics, teleconsultation, and e-prescription capabilities, these devices assist in enhancing access to healthcare for remote and rural populations. India is developing low-cost portable inventions to meet the demands of its sizable rural population.
c. **Telemedicine**: Telemedicine applies technology to remote monitoring, diagnosis, and instruction. It has made giving care in even the most remote regions possible while lowering provider and patient expenses.
d. **Digital and Social Connectivity**: Social media is widely used in India. Owing to enhanced telecom connectivity, 25% of the average person's time is spent on social networking sites. It facilitates the exchange of training and second opinions between physicians worldwide.
e. **Wearables**: These initially concentrated on diet and physical exercise when offered to the market. However, it is currently utilized to gauge fundamental health indicators.

### 12.5.1 Ayushman Bharat Digital Mission

Under the aegis of the NHP 2017, a committee led by Shri J. Satyanarayana was established to formulate an implementation framework. This committee's diligent efforts culminated in creating the NDHB in 2019. The NDHB serves as a comprehensive guide, delineating the essential components, and a concrete action plan for the widespread and meticulous integration of digital health solutions within India's healthcare landscape.

On August 15, 2020, a significant milestone was achieved when the Ministry of Health and Family Welfare unveiled the NDHM, firmly rooted in the principles set forth by the NDHB. This mission represents a pioneering initiative to establish a robust digital health ecosystem. Commencing as a pilot program within six union territories, the NDHM envisions the creation of a secure and interoperable healthcare system, facilitating the portability and accessibility of health records across both public and private healthcare institutions.

The NDHM revolves around three pivotal elements:

i. **Health ID**: Every resident has a Digital Health ID, serving as a secure repository for storing, accessing, and sharing their comprehensive health information. This Health ID may be linked to the individual's Aadhar and mobile number, allowing them to archive all health-related data and records that are readily accessible anywhere electronically. Sharing the credentials of this Health ID also empowers medical professionals with consent to access a patient's health information and historical records.
ii. **Health Facility Registry**: Organizations participating in the NDHM must register as healthcare providers. This inclusive category encompasses hospitals, clinics, diagnostic labs, imaging centers, and pharmacies. Once registered, healthcare facilities are bound to digitalize their systems in adherence

to the policies stipulated under the NDHM, rendering them legally binding. The compilation of registered healthcare facilities is maintained in a centralized repository, streamlining the process of digitally locating and accessing health services.

iii. **Healthcare Professional Registry**: This component facilitates traditional and modern healthcare professionals establishing an online presence. Healthcare providers who enlist in the NDHM's registry gain the capability to access patient records online and provide treatment through digital platforms.
iv. **Health Records**: Leveraging the health record mobile application system, individuals are empowered to actively contribute to and update their health-related data. Users can choose with whom they wish to share their data, whether healthcare practitioners, medical facilities, or other authorized entities.

In summary, the NDHM and its associated components are poised to revolutionize healthcare delivery in India by seamlessly integrating digital health solutions, enhancing accessibility, efficiency, and data security within the healthcare ecosystem.

## 12.6 RESTRICTIONS OF IMPLICATIONS OF DIGITALIZATION IN THE HEALTHCARE SECTOR

While technology has garnered significant popularity and widespread adoption across various industries, there remains a prevalent apprehension about the need for a certain degree of anonymity. This reluctance to fully embrace technological changes is common to all sectors, and the healthcare industry is no exception. In the context of the Indian healthcare ecosystem, several key data-related challenges must be surmounted:

**Data Privacy and Security**: The paramount concern in healthcare technology adoption is ensuring the privacy and security of sensitive patient data. Healthcare organizations need robust measures to protect patient information from unauthorized access, breaches, and cyber threats. Compliance with data protection regulations such as Health Insurance Portability and Accountability Act in the United States or European General Data Protection Regulation.

- **Interoperability**: The healthcare sector often grapples with the issue of interoperability, where different healthcare systems, devices, and applications struggle to share data and communicate seamlessly. Achieving interoperability is critical for holistic patient care and data accessibility.
- **Data Standardization**: Standardizing data formats and structures across various healthcare systems and providers is essential for efficient data sharing and analysis. Lack of standardization can lead to data fragmentation and hinder the development of comprehensive patient profiles.
- **Data Integration**: Many healthcare facilities have legacy systems that do not easily integrate with modern digital solutions. Bridging the gap between these systems and new technologies can be complex and resource-intensive.

- **Ethical Considerations**: Healthcare technology raises ethical questions about using patient data, consent, and potential biases in algorithms and decision-making processes. Striking the right balance between technological advancement and ethical considerations is a key challenge.
- **Data Quality and Accuracy**: Ensuring that healthcare data is accurate and reliable is crucial for clinical decision-making. Errors or inaccuracies in data can lead to incorrect diagnoses and treatment plans.
- **Health Information Exchange (HIE)**: A significant challenge lies in establishing effective HIE systems that facilitate the secure sharing of patient data across different healthcare providers and regions. This is particularly important for patient mobility and continuity of care.
- **Patient Engagement**: Encouraging patients to actively participate in their healthcare by sharing data, using digital health tools, and adhering to treatment plans is challenging. It requires overcoming issues related to health literacy, technology access, and patient trust in digital healthcare solutions.
- **Regulatory Compliance**: Keeping pace with evolving healthcare regulations and ensuring technology solutions adhere to these regulations is a continuous challenge. Failure to comply can result in legal and financial repercussions.
- **Resource Constraints**: Many healthcare institutions may lack the infrastructure and financial resources needed to adopt and maintain advanced healthcare technologies, especially in resource-constrained settings.
- **Patient Identification**: Locating or providing a Universal Health Identification can be challenging. The way that each hospital makes available the patient identification number varies.
- **Scattered and Fragmented Data**: Because patient's personal and health information is dispersed across various departments and information systems, it is difficult to gather and analyze collectively.
- **Data Ownership and Duplication**: The physician hesitates to hand over the patient's medical records. As a result, fresh data is generated when the same patient visits a different hospital.
- **Third-Party Involvement**: The insurance companies or other parties that help in the healthcare industry can often complicate the entire process, causing a delay in service and increasing overall costs for the industry.
- **Data Exchange**: Health policy planners find it challenging to gather, analyze, and communicate valuable health information more securely and smoothly due to the absence of unified healthcare industry standards, non-interoperable information systems, and a lack of platforms for health data exchange.

Various data-related challenges—including privacy, security, interoperability, standardization, ethics, data quality, patient engagement, regulatory compliance, and resource limitations—accompany technology adoption in the Indian healthcare industry. Overcoming these challenges is imperative for harnessing the full potential of technology to improve healthcare delivery and outcomes in India (Figure 12.5).

**FIGURE 12.5** Restrictions of implications of digitalization in the healthcare sector

## 12.7 CONCLUSION

Healthcare is among the most profoundly impacted e-services in the dynamic digital landscape. This article delves into critical facets such as improved information access, interpretation consistency, and heightened accountability. It underscores the significance, urgency, and challenges inherent in the ongoing digitalization process within the healthcare sector. At the core of this ambitious strategy lies the endeavor to efficiently reach an expansive patient population spread across vast geographic areas of India. Achieving this goal hinges on the meticulous digitalization of e-health information, executed with well-designed technologies. This strategic shift embodies a fundamental transformation in how healthcare is delivered and accessed in India.

The study explores the transformative power of modern technology within the Indian healthcare industry, transcending mere corporate and technical boundaries to usher in an era of personalized and integrated patient care. Furthermore, it enhances the management of healthcare outcomes over the long term. Recognizing that technology is just one facet of the broader digital healthcare landscape is crucial. Beyond adopting new tools and systems, digital healthcare introduces innovative approaches to healthcare challenges, enhances the patient experience, and accelerates the expansion of healthcare providers. This holistic transformation of healthcare operations promises to bolster institutions in delivering more effective services over extended periods, underscoring its profound significance for patients, healthcare professionals, and healthcare systems.

The global healthcare system stands on the brink of a profound transformation as the extensive digital health sector continues to develop, grow, and evolve. The impact of digital health transcends a mere trend; it symbolizes the future, promising enhanced access, efficiency, and effectiveness in healthcare delivery on a global scale.

## REFERENCES

1. Bardsley, Martin, Adam Steventon, and Garry Fothergill. *Untapped Potential: Investing in Health and Care Data Analytics*. London: Health Foundation, 2019.
2. Faujdar, Dharamjeet S., Sundeep Sahay, Tarundeep Singh, Manmeet Kaur, and Rajesh Kumar. "Field testing of a digital health information system for primary health care: A quasi-experimental study from India." *International Journal of Medical Informatics* 141 (2020): 104235.
3. Saini, C. P. "Difficulty in transition: Study on hindrances in online shopping for potential shoppers." *International Journal of Online Marketing (IJOM)* 10, no. 2 (2020): 17–29.
4. Faujdar, Dharamjeet S., Sundeep Sahay, Tarundeep Singh, Manmeet Kaur, and Rajesh Kumar. "Field testing of a digital health information system for primary health care: A quasi-experimental study from India." *International Journal of Medical Informatics* 141 (2020): 104235.
5. Lapão, Luís Velez. "The future of healthcare: The impact of digitalization on healthcare services performance." *The Internet and Health in Brazil: Challenges and Trends* (2019): 435–449.
6. Gjellebæk, Camilla, Ann Svensson, Catharina Bjørkquist, Nina Fladeby, and Kerstin Grundén. "Management challenges for future digitalization of healthcare services." *Futures* 124 (2020): 102636.
7. Moro Visconti, Roberto, and Donato Morea. "Healthcare digitalization and pay-for-performance incentives in smart hospital project financing." *International Journal of Environmental Research and Public Health* 17, no. 7 (2020): 2318.
8. McKee, M. M., Hauser, P. C., Champlin, S., Paasche-Orlow, M., & Wyse, K. "Using gaze tracking as a research tool in the deaf health literacy and access to health information project: Protocol for a multisite mixed methods study and preliminary results.", 1–12. *JMIR Research Protocols* 10, no. 9 (September 2021). https://doi.org/10.2196/26708.
9. Arora, Anmol. "Conceptualising artificial intelligence as a digital healthcare innovation: An introductory review." *Medical Devices: Evidence and Research*, no. 13 (2020): 223–230.
10. Saini, C. P., M. K. Nair, and K. T. Shankar. "Recycling and reuse of fashion: A way to sustainable environment." In A. Shrivastava, G. Jain, J. Paul (Eds.) *Circular Economy and Re-Commerce in the Fashion Industry* (pp. 47–53). IGI Global, 2020.
11. Kerras, Hayet, Jorge Luis Sánchez-Navarro, Erasmo Isidro López-Becerra, and Maria Dolores de-Miguel Gomez. "The impact of the gender digital divide on sustainable development: comparative analysis between the European Union and the Maghreb." *Sustainability* 12, no. 8 (2020): 3347.
12. Khalid, Ayesha, Ghulam Farid Malik, and Khalid Mahmood. "Sustainable development challenges in libraries: A systematic literature review (2000–2020)." *The Journal of Academic Librarianship* 47, no. 3 (2021): 102347.
13. Anjos, Luiz Antonio dos, Haroldo da Silva Ferreira, Nadya Helena Alves-Santos, Maiara Brusco de Freitas, Cristiano Siqueira Boccolini, Elisa Maria de Aquino Lacerda, Inês Rugani Ribeiro de Castro et al. "Methodological aspects of the anthropometric assessment in the Brazilian National Survey on Child Nutrition (ENANI-2019): A population-based household survey." *Cadernos de Saúde Pública* 37 (2021): e00293320.

14. Chandra, Amitabh, Jonathan Holmes, and Jonathan Skinner. *Is this Time Different? The Slowdown in Healthcare Spending*. No. w19700. National Bureau of Economic Research, 2013.
15. Ancker, Jessica S., Lisa M. Kern, Alison Edwards, Sarah Nosal, Daniel M. Stein, Diane Hauser, Rainu Kaushal, and with the HITEC Investigators. "Associations between healthcare quality and use of electronic health record functions in ambulatory care." *Journal of the American Medical Informatics Association* 22, no. 4 (2015): 864–871.
16. Arora, Vishal S., Martin McKee, and David Stuckler. "Google trends: Opportunities and limitations in health and health policy research." *Health Policy* 123, no. 3 (2019): 338–341.

# 13 Unleashing the Potential of Machine Learning

## *Revolutionizing Loan Forecasting Using Logistic Regression*

*Pranshu Mishra, Nyasha Saurabh, Rishav Katoch, and Richa Sharma*

## 13.1 INTRODUCTION

In today's world, we are all aware that one needs a certain amount of capital to pursue any endeavor or goal in life. In our lives we are often presented with various opportunities. Whether it's some revolutionary business idea or a chance to study in one of the most esteemed educational institutions. But more often than not people have to lose out on these opportunities due to lack of financial resources. They may try to apply for loans and grants but the backbreaking job of loan applications can be tiresome. To smooth out the process, we have created a loan prediction model with a model accuracy score of 87%, which is quite high compared to other machine learning (ML) models present. We have integrated this model to a website-based interface to make it more user-friendly. Whether it be a pursuit of higher education, a dream for one's own home or a top-of-the-line car, or maybe even a small business, we may be cut short due to a lack of capital or resources. Think about how many small business start-ups didn't achieve their potential due to a lack of resources or a bright student with an impeccable track record being unable to get money for further higher education. Think of how much potential remains untapped in our world due to a lack of resources and all the potential Steve Jobs and Ratan Tata who were unable to pop off due to such financial constraints. To cover this lack of capital, a certain amount of it can be taken out as loans from various financial institutions but many of us are aware of the grueling process and the difficulty associated with the process of approval of loans. Financial institutions employ loan prediction forecast models to determine if a loan application is likely to repay the loan or default on payments [1]. Typically, ML techniques are used to create models of this kind. These algorithms examine past data on loan applications and outcomes to spot trends and forecast future behavior.

DOI: 10.1201/9781032700953-15

The loan prediction forecast model is an important tool for financial institutions seeking to make data-driven lending decisions and minimize their risk exposure. In general, loan prediction forecast models can offer lenders several advantages.

## 13.2 LOGISTIC REGRESSION

To investigate and analyze the link between a binary-dependent variable and one or more independent variables, logistic regression statistical analysis is used. In the context of a loan prediction forecast model, logistic regression can be used to forecast how likely a borrower is to default on their loan or not based on certain attributes of the given borrower, such as their credit score, income, age, and so on [1, 2].

1. The volume of the unbiased factors can be utilized to understand the effects of each factor on the likelihood of default in logistic regression models because they are relatively simple to read.
2. These models are computationally efficient and are able to handle a great number of independent variables [3].
3. Logistic regression models can be used with numerical and categorical independent variables, making incorporating a wider range of borrower characteristics easier.

## 13.3 METHODOLOGY

The primary goal is to introduce the clear objective of the loan prediction model using supervised learning algorithms. The presented loan prediction system is an ML application integrated with Python Flask using logistic regression for better accuracy. To begin with, in this model, we need to load libraries such as pandas, numpy, matplotlib, and seaborn. After that, we have to read the data-set by pd. read. Train and test data-set have all columns same except the target column that is "Loan Status". The training data-set includes 13 columns and 614 rows, while the test data-set includes 12 columns and 367 rows as target columns are not inserted into the test data-set. Also, there are two columns available that is categorical columns (values in the form of words such as male/female, graduate/not graduate) and numerical columns (values in the form of numbers such as loan ID, applicant income, co-applicant income) [4]. After getting sufficient information about data-set, we will handle the missing values of columns, we will fill the null values of categorical columns by the mode of that column, and for numerical columns, we will fill the null values by the mean of that column. Now, we will use mapping functions to map the categorical columns' value into a numeric, as ML understands only numeric values. We will map "Male" as "1" and "Female" as "0" [5]. Splitting the data-set into two portions, that is, train and test, so that we can perform exploratory data analysis. The analysis contains two observations, that is, uni-variate and bi-variate. Seaborn is imported for graphical analysis. For some of the analyses, the count of males is more than females, a maximum number of applicant has one child, a maximum number of applicant is married, etc. After a perfect analysis, we found that in numerical columns, maximum and minimum difference is huge; now, we will transform the

```
# logistic regression
from sklearn.linear_model import LogisticRegression
model3 = LogisticRegression()
model3.fit(x_train, y_train)
print("Accuracy is", model3.score(x_test, y_test)*100)

Accuracy is 87.8048780487805
```

**FIGURE 13.1** Accuracy score

data. To transform the data into numerical columns, we will take log values of all entries that is log value of applicant income, log value of co-applicant income, etc. We will store the feature value in X and target value in Y. We will drop the loan status column from X because it is the target value. Now after correcting all the fields of data-set, we will split the data-set into two parts that are train and test data-set. We have a (80:20) split in data-set.

Now we will use different ML algorithms for training the model, such as logistic regression, support vector machine (SVM) random forest classifier, decision tree classifier. We will import all algorithms from sklearn and we will test the accuracy of different algorithms. We will create an instance of logistic regression and then we will call the model to check the accuracy score [6]. Among all the algorithms used to train the model, we found that logistic regression leads testimony data with an efficiency score of 85.20%. In order to obtain a model with optimal performance, we have done hyper tuning to improve accuracy by 87.80% using RandomizedSearchCV.

The logistic regression model's accuracy score increased from 85.20% to 87.80% after being hyper-tuned (Figure 13.1). At last, we will save the model in.pkl form.

We will call load the saved.pkl file into our website. Our website will gather different values that are present in data-set, that is, gender, number of children, applicant income, credit score, etc.

The prediction of whether a loan is accepted or not is through a pop-up message that is organized in the app.py. To execute the website, we have to run it on vs-code, and at last, to end the application click CTRL+C (Figure 13.2).

### 13.3.1 Proposed-Model Solution Process

Step 1: Start
Step 2: Import different Libraries
Step 3: Read the data-set
Step 4: Take the information about data-set
Step 5: Handle the missing values in data-set
Step 6: Map the categorical columns into numeric forms
Step 7: Perform both statistical and graphical data analyses
Step 8: Perform Feature Scaling
Step 9: Store feature, response matrix in X, Y

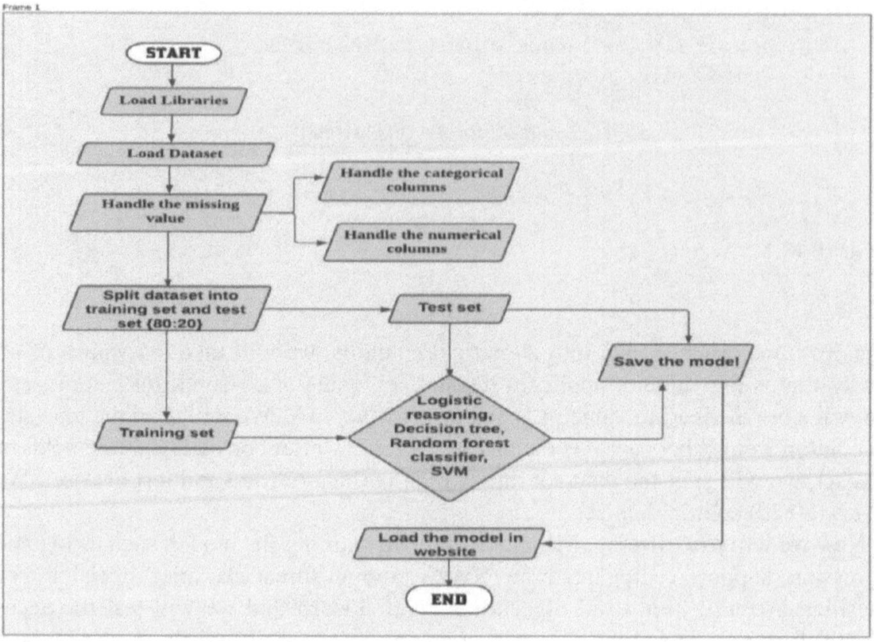


**FIGURE 13.2** Flow diagram

Step 10: Split the data-set into train and test
Step 11: Perform different algorithms to train the model
Step 12: Perform hyperparameter tuning to achieve better accuracy
Step 13: Save the created model file
Step 14: Introduce the model file to the webpage
Step 15: End

## 13.4 CONFIGURATION FRAMEWORK

OS – Windows 10 or more
Processor – Ryzen 7, 4800H
GPU – RTX3050
RAM – 15.8 GB
Compilers – Jupyter Notebook, Anaconda
Programming Language – Python 3.11.1

## 13.5 DIFFERENT ALGORITHMS IN MACHINE LEARNING MODEL TRAINING

### 13.5.1 Decision Tree

A prominent machine learning strategy that can be used to develop a loan prediction model is decision trees. Using the input features (such as the borrower's income, credit score, and employment history), a decision tree in this situation solves the loan prediction problem by breaking it down into a series of smaller, easier decisions. The algorithm examines the available information at each stage of the decision tree to find which one is most instructive for predicting whether or not to approve the loan (Figures 13.3–13.5). The data is then divided according to that feature, and the procedure is continued with the new subgroups until a decision is determined [7].

### 13.5.2 Random Forest Classifier

Another machine learning approach that can be used to create a loan prediction model is the Random Forest Classifier. To generate a final prediction, various predictions from every decision tree are combined into a single prediction, creating a "forest" of decision trees. The Random Forest builds each decision tree using a randomly chosen portion of the available features and a randomly selected subset of the training data (Figures 13.3–13.5). The Random Forest Classifier runs a fresh loan application through each decision tree in the forest to generate predictions for it, which are then combined to get a final forecast [7].

### 13.5.3 Logistic Regression

A popular machine learning approach called logistic regression is utilized to create a loan prediction model. Based on the values of the input features (such as the borrower's income, credit score, and employment history), it calculates the likelihood of loan approval (Figures 13.3–13.5). The approach builds a model that estimates the probability of approval for a specific set of input features in logistic regression, where the output is a binary variable (approved or denied loan). By a logistic function being fitted to the training data-set, the model learns the correlation between the input features and the likelihood of acceptance [7].

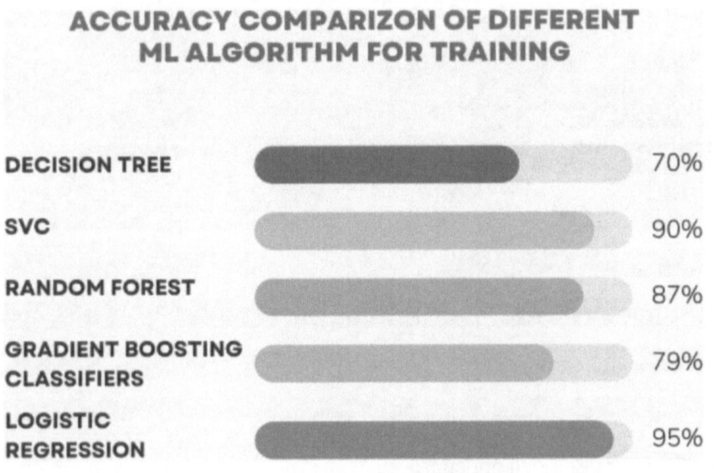


FIGURE 13.3 Accuracy evaluation of several algorithms

### 13.5.4 Support Vector Machine

The technical phrase is SVM. The SVM algorithm chooses the optimal hyper-plane for categorizing the data by maximizing the distance between the hyper-plane and the closest data points of each class (Figures 13.3–13.5). SVM can also use kernel functions to transport the information into a more complex space where a linear hyper-plane can make it easier to distinguish between the classes. SVM is a powerful technique that can handle complex data and is commonly applied to a variety of tasks, such as image classification, text classification, and bio-informatics [7].

**FIGURE 13.4** Output – Loan approved

**FIGURE 13.5** Output – Loan not approved

## 13.6 RESULT AND ANALYSIS

Through the use of this web-integrated, ML-based loan prediction model, we aim to smooth out the kinks that may arise in the process of loan approval, thus making it more easy for people to get their loans approved. We will load certain libraries and then we will read the data-set into which we will handle the null values; after handling the null values, we will map the categorical columns into numeric forms. To improve the maximum and minimum value difference, we will use feature scaling [8, 9]. After that, we will split the data-set into test and train data and we will perform different model training algorithms to achieve the best accuracy score. After that, we will save the model.

The saved model will be loaded into the website where the user has to give basic details such as gender, graduation, marital status, and income. Based on the user's information, this model will predict whether loan is approved or not. A pop-up message will be displayed that your loan is rejected or accepted.

In this chapter, there is a highly efficient (87%) loan prediction model and its comparison with different algorithms is discussed.

- We have discussed logistic regression and its uses, features, etc.
- We have compared different number of algorithms and concluded that logistic regression was the most accurate.
- We have outlined the configuration parameter descriptions.
- Different model training algorithms are compared (comparison result is based on model training).

## 13.7 FUTURE SCOPE

We will work on a variety of plans, including:

- By adding new data and utilizing more sophisticated machine learning methods, the accuracy of the model can be further increased. Better forecasting and decision-making in the loan sector may result from this.
- The website can work with banks in order to help them in prediction. This can aid lenders in making wise lending decisions and better risk-management decisions.
- Website can be improved in many forms so that the loan process could become more effective as a result.
- Loan prediction forecast models can be used to identify and stop fraud in the lending sector.

## 13.8 CONCLUSION

The foremost motive of the chapter is to classify and inspect the nature of the loan candidates. Loan candidates are individuals or businessmen who apply for loans. From a proper study of the data and limitations of the banks, it can be concluded that this model is highly efficient with an accuracy of 87.80% [10]. With this accuracy rate, this system is highly capable of predicting outcomes. This system was intelligent enough to forecast whether the candidate will be eligible for the loan or not. Hyper tuning increased accuracy. Logistic regression performed superior to every other ensemble model. This loan prediction system can assist banks as this model is associated with websites, which will make a banker's job easy.

## REFERENCES

1. Prakash, C., Yadav, R., & Kadyan, S. (2021). Effect of the price drop on customer's perceived evaluation across selected product categories. *Journal of Revenue and Pricing Management*, 20, 204–210.
2. Saini, C. P., Nair, M. K., & Shankar, K. T. (2020). Recycling and reuse of fashion: A way to sustainable environment. In C. P. Saini, M. K. Nair, & K. T. Shankar (Eds.), *Circular Economy and Re-Commerce in the Fashion Industry* (pp. 47–53). IGI Global.
3. Prakash, C., Yadav, R., Singh, A., & Yadav, K. (2021). Modeling the entrepreneurial Intention among business Students. *Pacific Business Review International*, 14(3).
4. Prakash, C., Yadav, R., Dangi, A., & Singh, A. (2023). Factor analysis approach to study mobile applications' characteristics and consumers' attitudes. In P. Chatterjee, D. Pamucar, M. Yazdani, & D. Panchal (Eds.), *Computational Intelligence for Engineering and Management Applications: Select Proceedings of CIEMA 2022* (pp. 783–798). Singapore: Springer Nature Singapore.
5. Saini, C. P. (2020). Difficulty in transition: Study on hindrances in online shopping for potential shoppers. *International Journal of Online Marketing (IJOM),* 10(2), 17–29.
6. Chaabene, Wassim Ben, Flah Majdi, & Nehdi, Moncef L. (2020). Machine learning prediction of mechanical properties of concrete: Critical review. *Construction and Building Materials,* 260, 119889.
7. Singh, Pooja, Bahubal, Lipsa Dev, Rani, Sangeeta, & Kumar, Ajay. (2023). AI/ML revolutionizing social media and business: A comprehensive study. In A. Kumar, S. Rani, S. Rathee, & S. Bhatia (Eds.), *Security and Risk Analysis for Intelligent Cloud Computing* (pp. 252–279). CRC Press.
8. Singh, Amanpreet, Thakur, Narina, & Sharma, Aakanksha. (2016). A review of supervised machine learning algorithms. In *2016 3rd International Conference on Computing for Sustainable Global Development (INDIACom)*, pp. 1310–1315. Ieee.
9. Ayyadevara, V. Kishore. *Pro Machine Learning Algorithms.* Berkeley, CA: Apress, 2018, 283–297.
10. Osisanwo, F. Y., Akinsola, J. E. T., Awodele, O., Hinmikaiye, J. O., Olakanmi, O., & Akinjobi, J. (2017). Supervised machine learning algorithms: Classification and comparison. *International Journal of Computer Trends and Technology (IJCTT)* 48(3), 128–138.

# Index

Note: **Bold** page numbers refer to tables; *italic* page numbers refer to figures